4th International Conference on Design and Manufacturing Engineering 2019

IOP Conference Series: Materials Science and Engineering
Volume 627

Geelong, Australia
15 – 17 July 2019

ISBN: 978-1-5108-9831-8
ISSN: 1757-8981

Printed from e-media with permission by:

Curran Associates, Inc.
57 Morehouse Lane
Red Hook, NY 12571

Institute of Physics
Dirac House, Temple Back
Bristol BS1 6BE UK

Phone: 44 1 17 929 7481
Fax: 44 1 17 920 0979

techtracking@iop.org

Additional copies of this publication are available from:

Curran Associates, Inc.
57 Morehouse Lane
Red Hook, NY 12571 USA
Phone: 845-758-0400
Fax: 845-758-2633
Email: curran@proceedings.com
Web: www.proceedings.com

Table of contents

Volume 627

4th International Conference on Design and Manufacturing Engineering

15–17 July 2019, Geelong, Australia

Accepted papers received: 27 August 2019
Published online: 18 October 2019

Preface

Papers

Manufacturing Technology

IOP Conf. Series: Materials Science and Engineering **627** (2019) 011001 doi:10.1088/1757-899X/627/1/011001

Preface

I was delighted to serve as the local chair for the 4th International Conference on Design and Manufacturing. The event was held at Deakin University's Institute for Frontier Materials in Geelong, Australia. The conference was also supported by the ARC Research Hub for Future Fibres based at Deakin University.

Deakin University is a large comprehensive university, with over 60,000 course enrolments and a fast growing reputation for research that makes a real difference. The Institute for Frontier Materials is the largest research institute at Deakin, and currently enjoys Australia's highest research excellence ranking score of 5 in both materials engineering and manufacturing engineering, areas closely related to this conference.

Design and manufacturing go hand in hand, and the design aspect will become more and more important in future manufacturing. This is driven by the global trend for sustainability and circularity.

The conference provided a great forum to exchange recent research activities around the globe and I hope you will enjoy the selected papers from this conference. I also look forward to seeing you at a future conference.

Yours sincerely,

Professor Xungai Wang

IOP Conf. Series: Materials Science and Engineering **627** (2019) 011002 doi:10.1088/1757-899X/627/1/011002

Conference Committee

Conference Committee Chairs

Wenlong Cheng, Monash University, Australia

Yonggang Zhu, Harbin Institute of Technology - Shenzhen, China

Conference Local-Chair

Xungai Wang, Deakin University, Australia

Program Committee Chairs

Zhengyi Jiang, University of Wollongong, Australia

Xin-Ping Zhang, South China University of Technology, China

International Technical Committees

Xiaodong Huang, Swinburne University of Technology, Australia

Laichang Zhang, Edith Cowan University, Australia

Qiaoliang Bao, Monash University, Australia

Mehmet Yetmez, Bulent Ecevit University, Turkey

Mustafa Taşyürek, Selcuk University, Turkey

Masita Mohammad, Universiti Kebangsaan Malaysia, Malaysia

Rosli Mohamad Zain, Universiti Teknologi Malaysia, Malaysia

Md. Mustafizur Rahman, Universiti Malaysia Pahang, Malaysia

Devarajan Ramasamy, Universiti Malaysia Pahang, Malaysia

Winda Nur Cahyo, Islamic University of Indonesia, Indonesia

Md. Anayet U Patwari, Islamic University of Technology, Bangladesh

Syed Husain Imran Jaffery, National University of Sciences and Technology (NUST), Pakistan

Mushtaq Khan, National University of Sciences and Technology (NUST), Pakistan

Mahsa Baghban Salehi, Chemistry and Chemical Engineering Research Center of Iran (CCERCI), Iran

Sriram Venkatesh, University College of Engineering, India

Vishwas Deshpande, Ramdeobaba College of Engineering & Management, India

Bhaskar Pal, Presidency University, Bangalore, Karnataka, India

Peer review statement

All papers published in this volume of *IOP Conference Series: Materials Science and Engineering* have been peer reviewed through processes administered by the proceedings Editors. Reviews were conducted by expert referees to the professional and scientific standards expected of a proceedings journal published by IOP Publishing.

IOP Conf. Series: Materials Science and Engineering **627** (2019) 012001 doi:10.1088/1757-899X/627/1/012001

Measuring effectiveness and efficiency of a new product development project using performance matrix – The PROTON case

H Boejang[1,*], M Z Hassan[1], T Mokudai[2], H Ariff[3] and S Esa[4]

[1]*Fakulti Teknologi Kejuruteraan Mekanikal dan Pembuatan, Universiti Teknikal Malaysia Melaka, Hang Tuah Jaya, 76100 Durian Tunggal, Melaka, Malaysia*
[2]*Faculty of Economics, Kyushu University, 744 Motooka Nishi-ku, Fukuoka 819-0395, Japan*
[3]*Fakulti Kejuruteraan Pembuatan, Universiti Teknikal Malaysia Melaka, Hang Tuah Jaya, 76100 Durian Tunggal, Melaka, Malaysia*
[4]*Perusahaan Otomobil Nasional Sdn. Bhd. (PROTON), Persiaran Kuala Selangor, Seksyen 26, 40400 Shah Alam, Selangor Darul Ehsan, Malaysia*
**Correspondance author: hambaliboejang@utem.edu.my*

Abstract. This paper presents a study on the development of a performance matrix to measure the new product development project performance for a small car producer in Malaysia. This study was carried out through open-ended interview sessions, involving a couple of project managers, a new product introduction manager and senior product planning manager and engineers. The questionnaires were developed based on the literatures and pre-tested prior the interview sessions. The feedbacks were used to identify the most important internal and external performance outcomes that crucial to the car producer. The development lead-time represents the efficiency, while the market share represents the effectiveness of the product development process. The performance matrix was successfully developed and the relation between the development lead time and market share is asymmetric. This performance matrix is best used for comparing NPD performances between plan and actual of a developed project, as well as between actual performances of a previous project and a new project that are having a similar project scopes and contents. The application of the performance matrix through a case study indicates that the success or failure status of the new product development project can be determined. The performance matrix is found to be viable and suitable for this indigenous car producer.

1. Introduction

This study focuses only on the completed development projects of Perusahaan Otomobil Nasional Sdn. Bhd. (PROTON) that were launched between the year of 2000 until the mid of 2017, before the acquisition of 49.9% stake in Proton by Zhejiang Geely Holding Group. During this time, PROTON managed to increase the frequency of the New Product Introduction (NPI) by more than 100% [1]. Prior 2000, most of the New Product Development (NPD) projects were only involved the lower degree of enhancement and facelift types of projects due to the constraints of limited resources i.e. product development experts and capability, monetary investment, and man powers. However, circa 2000 the establishment of a well-defined integrated product development practice and the continuous improvement of this formal practice enabled PROTON to introduce several models for the local market either through own development or collaboration with other producers. Although, the challenges of limited pool of resources were not holistically overcome; with appropriate business strategies PROTON successfully achieved its performance targets i.e. shorter development time and lower development cost.

This study is carried out based on the request from the respondents of having a simple and straight forward decision-making tool for measuring NPD performance. Product development project performance has been studied by many scholars based on different theoretical perspectives together with different parameters or variables [2-4]. An NPD performance is a significant subject matter in engineering management field. The performance of an NPD project may be considered as a success at the project or program-level, but the product may ends-up as failure right after it enters the market. In addition, a set of specific performance measures may not work well for other product development projects in different industries at various international locations. These two examples depict how an NPD performance must be managed astutely. The selection of appropriate performance measures is so crucial for product development practitioners in order to completely understand what the best parameters to be used to measure the NPD performance. As such, for this study the development of the performance matrix for measuring Proton's NPD performance depends on the feedbacks obtained from the respondents during the interview sessions in which the efficiency and the effectiveness of the NPD are considered and become the main interest.

2. Literature review

The performance of an NPD is closely related to the project or program goals set by the product producers which could be classified as the internal or external outcomes. The NPD performance is influenced by many factors such as project content, product complexity, development team, and knowledge management for problem solving. Lee and Markham [5] emphasise the importance of having a flexible NPD that suits specific environments in order to increase new product success. Thus, as an established car producer Proton adopt its own formal product development practices known as new product introduction (NPI) that meets its own business conditions.

2.1 Internal and External Project Goals.

The success and failure of an NPD have been the interest of practitioners and researchers from different backgrounds and industries. The NPD performance measures consist of many determinants such as development speed, market share, profitability, and cost [6]. Griffin and Page [7] provide an interim report that represents the findings of a Product Development and Management Association (PDMA) workforce on dimensions used for measuring success and failure of an NPD. These dimensions, in general include customer acceptance measures, financial performance, product-level, and firm-level. Furthermore, they can be classified as the internal or external project goals i.e. unit cost, lead time, unit sale, breakeven, and profit.

2.1.1 Development lead time. The development lead time is considered as an internal project goal. Another examples of this types of goals are development cost, product performance, meeting quality goals, and launched on time [8-10]. Clark and Fujimoto [11] define the development lead time as the time in month between the initial stage of a development project and the introduction of a first model to the market. Many scholars have proved that building product quickly and on time is beneficial to the manufacturers [12-14]. In short, development lead time is a dimension that tells us how project execution adheres to the project schedule.

2.1.2 Unit sale and market share. The unit sale and market share are known to be the external project goal since it reflects the product performance in the market. Some of the external project goals are commonly known as customer acceptance and satisfaction, return on sales, breakeven and breakeven point [15]. According to Page [16] sales and profits are measures that quantify the impact of development program on the profit and loss statement of a firm and tell the effect of the NPD activities on financial performance. Therefore, an assumption was made that the higher the unit sale and market share, the higher the effectiveness of the NPD. In this case, unit sale refers to the amount of a newly introduced product being sold within a specific period i.e. monthly or yearly basis. On the other hand, the market share refers to a portion of a market controlled by a product.

2.1.3 The efficiency and effectiveness of the NPD. The development lead time is considered as the efficiency, and the monthly or yearly unit sales as well as the market share are representing the effectiveness of the NPD. These three dimensions are considered based on the facts that shorter development cycle will not have any effect over product success, if the new product does not meet the requirements of the customers [17], and product success is best assessed by combining market share, profitability and customer satisfaction [13]. Therefore, this infers that the efficiency and effectiveness of an NPD are influenced by the development lead time, the unit sale and the market share of a product. With reference to Wheelwright and Clark [17], a conclusion is made: the effectiveness based on the market share is more significant than the unit sales. The market share tells the impact of the newly introduced product in terms of relative competitiveness i.e. a competitive product is expected to have a decent or positive sale volume even the economy situation is in down trend. Of course, both dimensions for measuring effectiveness are more important than the development lead time. Thus, it is believed that the efficiency and effectiveness of the NPD are asymmetric. Moreover, since PROTON is known to be as the 'people car' due to lower or affordable price offered to customers mainly at the bottom of the pyramid, an increase in sale must be one of its main concerns. Unlike the previous research works, the development of a decision-making tool to determine the NPD performance is the main concern of this study.

3. Research Approach

Figure 1 illustrates the process of the systematic research review in this study. It involves massive effort, pertaining to the new product development process and project performance from various perspectives. Information was gathered and reviewed across several databases i.e. SCOPUS, Google Scholar, Springer Link, ScienceDirect, and IEEE Explore through a systematic research review method. The documents were first extracted based on years of publication from the mid-80s' to 2018. Second, they must be available in full-text in English. Third, the publications had to include all the peer-reviewed articles and relevant to new product development. Forth, the publication must consider reliable writers who have had more than five publications related to the topic and must have more than five citations after 2000. And finally, the publications were categorized into specific folders in Mendeley database, and duplicate publications were deleted.

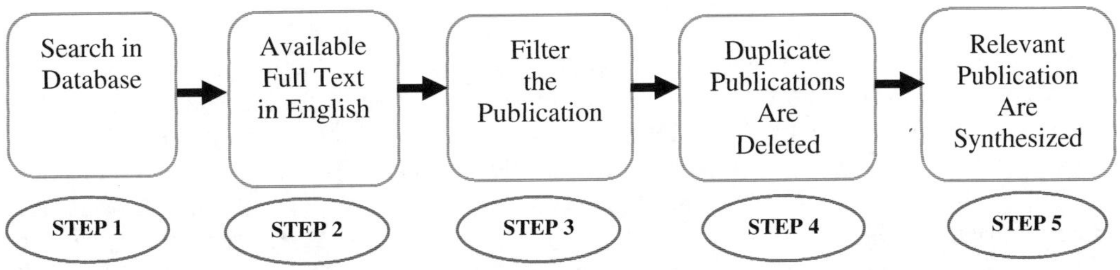

Figure 1. The systematic literature reviews.

Initially, a draft version of the questionnaire was sent off via email to the contact person for pre-testing purpose. The questionnaire was cross-checked against the literature and with the respondents. Later, the questionnaire was fine-tuned; a face-to-face interview was planned and scheduled. The interview session was conducted at the project management office in the car producer's assembly plant a few months later. The participations involved were a senior manager from the planning division and a senior executive from the NPI unit, along with two project managers from two different vehicle programs, and four development team members: two from each of the vehicle programs. Even though, the number of the participants is small, most of them have involved in many development projects. The interview was an open-ended semi-structured where the discussed topics cover without limitation the formal NPD process and project, the project characteristics, and finally the NPD performance. To suit the requirements from the respondents, the unit sale, market share and the development lead time were used to measure the effectiveness and the efficiency of Proton's NPD performance. Although, the development cost is another significant internal project goal for Proton, due to confidential issue it was ignored. This leads to a decision to exclude all financial performance measures as well. In fact, in Malaysia the car companies are not allowed to let the public know about the sales data by the Malaysian Competition Commissioner (MyCC). The development lead time reflects the efficiency of the NPD project while the annual sales volume and the market share reflect the effectiveness of the NPD project.

Finally, to compare those performance outcomes a simple yet straight forward performance matrix was developed in which both outcomes were measured and compared. The comparison of the performance measures can be made for a single project i.e. plan versus actual, and for similar types of development projects i.e. new platform versus new platform, full model change versus full model change that considerably involve design contents and project scopes of about the same weight. This performance matrix is known as Boejang Performance Matrix (BPM). The outcomes from the evaluation determine the performance of a specific NPD project. Some specific levels of performance which reflect the efficiency and the effectiveness of the NPD i.e. lower than before, same as before, better than before were assigned to each of the boxes in the matrix. A case study was used to illustrate the application of the BPM.

4. Results and Discussion

Data from the literature and the interviews were used to establish the performance matrix. Since both performance goals are asymmetric, the following Table 1 were developed. An NPD effort is failed when the efficiency and the effectiveness met the following conditions:

i. Lower efficiency and lower effectiveness.
ii. Lower efficiency and same effectiveness.
iii. Same efficiency and same effectiveness
iv. Same efficiency, and lower effectiveness and higher efficiency and lower effectiveness.

Meanwhile, an NPD is considered to meet its performance success when these conditions are occurred:

i. Higher efficiency and same effectiveness (partially successful).
ii. Same efficiency and higher effectiveness (successful).
iii. Higher efficiency and higher effectiveness (very successful).

There is only one condition where the NPD performance is classified as "difficult to judge" when the efficiency is lower, and the effectiveness is higher. The levels: lower than before, same, and higher than before for the efficiency and effectiveness are based on comparison between actual against the planned development lead times (month) and unit sales (volume) for a single project. Furthermore, it also refers to the comparison in terms of efficiency and effectiveness of the previous and new development projects. The 3rd generation of PROTON *Saga (New Saga)* which was launched in September 2016 was used in this case study. It was developed to replace the 2nd model (*Old Saga*). Table 2 shows some data of the developed models. These data were obtained from a couple of websites especially from the Malaysia Automotive Association (MAA) and Astro AWANI online news as well as from the respondents. The total numbers of passenger vehicles sold in 2017 is 70,991 units, and 80% is contributed by the selling of the *New Saga, Persona* and *Ertiga*, and according to the Vice President, Sales and Marketing of Proton the total number of units sold for *Persona* is 19,510 units, and for *Ertiga* is 6,091 units [18, 19]. 80% out of 70,991 is equal to 56,793 units. By deducting the units sold of *Persona* and *Ertiga*, the total numbers of unit sold for the *New Saga* is 31,200 units or at an average of 2,600 units/month.

Table 1. Boejang Performance Matrix

		Efficiency		
		Lower than before	Same	Higher than before
EFFECTIVENESS	Lower than before	New NPD failed	New NPD failed	New NPD failed
	Same	New NPD failed	No improvement (NPD failed)	Partially successful
	Higher than before	Difficult to judge	Successful	Very successful

Table 2. Development time and unit sold of *PROTON Saga* Model.

Model	Year	Development lead time(month)		Average Unit sale/month	
		Plan	Actual	Plan	Actual
New Saga	2017	24	16	3,000	2,600
Old Saga	2008	24	18	3,000	5,544

It is found that the development lead time for the *New Saga* is reduced by 33.3% from the planned development lead time. The eight months' reduction in the development lead time shows that the NPD is efficient. On contrary, the actual unit sale of the model in 2017 is 13.3% lower than the planned volume. Meaning, the efficiency of the NPD is improved, but not the effectiveness of the NPD where the average of the actual unit sale of 2,600 units/month is lower than the target unit sale by 400 units. Therefore, the NPD is considered to be a failure as it has a higher efficiency, but lower effectiveness. The BPM can also be used to judge on the NPD performance between the *Old Saga* also known as *Saga-BLM* which was released in 2008 and its successor which is the new model after a year both products being released to the market. However, Table 2 only shows the average monthly sales unit, since the exact monthly sales data for both models are not available due to restriction imposed by the MyCC.

From the Table 2, the actual development lead time for the *New Saga* is two months lower than the *Old Saga* which represented by 11.1% reduction by comparison. Thus, its efficiency is higher than before. For the unit sale there is a decrease in effectiveness as the average monthly unit sale for the new model is 2,944 units less or 53.1% lower than the average monthly unit sale of the old one. As such, with reference to the BPM the NPD performance of the New Saga is a failure when compared with the old model.

Table 3. Development time and unit sold of *Proton Saga*.

Project	Market Report (year)	TIV (units)	Total PV	Total Unit Sale	Total Market Share (%)	Market share (%)
New Saga	2017	514679	70991	31200	13.8	43.9
Old Saga	2008	525770	141958	66533	27.1	46.9

Table 3 above shows some findings of both products for effectiveness test. The market report from MAA tells that the Total Industry Vehicle (TIV) in 2017 for passenger vehicle (PV) is equal to 514,679 units where 70,991 units are PROTON. Thus, PROTON controls 13.8% of the market share of the passenger car [18]. The *New Saga* total unit sales in 2017 is 31,200 units in which it represents 43.9% of the total PV of PROTON. Meanwhile, in 2008 the TIV of the PV is 525,770 units in which 141,958 units are PROTON. This is about 27.1% of the market share of the PV in that year. As the total unit sale of the *Old Saga* is 66,533, it represents 46.9% of total PV sold by PROTON in 2008

[20]. From those figures it is very clear that the market share controlled by PROTON in 2008 is better than in 2017. The unit sale of the *Old Saga* is about doubled the unit sale of the *New Saga*. With reference to the BPM, it is found that the NPD performance of the *New Saga* is a failure as well, although the efficiency is higher, the effectiveness is lower than before. However, from the observation it is also found that both the *New Saga* and *Old Saga* have significant impacts on the market share as both contribute 43.9% and 46.9% of the total number of the PV of PROTON relatively. As such, it can be said that the low market shares of PROTON's passenger vehicles in 2017 (13.8%) may be due to some other reasons i.e. promotion, macro-economic scenario, competitors' strategies that are differed from the 2008. It also suggests that PROTON must address those possible external factors that affected the market share in order to have a better insight into its NPD performance. Moreover, for PROTON to have a better picture of its NPD project performance it is best to use the actual data of the monthly unit sale rather than the average monthly unit sale used in the case study since the outcomes would enable the practitioner to understand the actual comparative improvement between the NPD projects, and to have a better control in the product life cycle of the new and old models within certain periods of times.

5. Conclusions

Based on the findings and case study it is concluded that the BPM is reliable and suitable to be used as a decision-making tool that tells the status of an NPD performance. The BPM is applicable in determining the success and failure of an NPD by comparing the efficiency and effectiveness of the product development projects or programs. The measurement of the effectiveness of an NPD project in terms of market share gives better picture on the NPD performance rather than the unit sale. From the case study, the application of the BPM is found to be feasible on both cases. The adoption of a well-defined NPD is crucial, and the selection of types of project is very important for comparing the NPD performance when using the BPM. The efficiency of PROTON's NPD has improved over time as the development lead time is getting better. Further investigation must be carried out in order to test the reliability of the BPM. The respondents are willing to use BPM as decision making tool for the future development projects.

References

[1] Boejang H, Ariff H, Hassan M Z, Esa S and Rauterberg M, An exploration on new product development process of Malaysian small-sized automaker, *J. Adv. Manuf. Technol.*, 11, no. 2, 2017.
[2] Cooper R, "How new product strategies impact on performance, *J. Prod. Innov. Manag.*, 1, no. 1, pp. 5–18, 1984.
[3] Awwad A and Akroush D M N, New product development performance success measures: an exploratory research, *EuroMed J. Bus.*, 11, no. 1, pp. 2–29, 2016.
[4] Pattikawa L H, Verwaal E and Commandeur H R, "Understanding new product project performance, *Eur. J. Mark.*, 40, no. 11–12, pp. 1178–93, 2006.
[5] Markham S K and Lee H, Product development and management association's 2012 comparative performance assessment study, *J. Prod. Innov. Manag.*, 30, no. 3, pp. 408–29, 2013.
[6] Lewis M A, Success, failure and organisational competence: a case study of the new product development process, *J. Eng. Technol. Manag.*, 18, no. 2, pp. 185–206, 2001.
[7] Griffin A and Page A L, An Interim Report on Measuring Product Development Success and Failure, pp. 291–308, 1993.
[8] Swink M, Completing projects on-time: how project acceleration affects new product development, *J. Eng. Technol. Manag.*, 20, no. 4, pp. 319–44, 2003.
[9] Yang L R, Implementation of project strategy to improve new product development performance, *Int. J. Proj. Manag.*, 30, no. 7, pp. 760–70, 2012.
[10] Swink M, Talluri S and Pandejpong T, Faster, better, cheaper: A study of NPD project efficiency and performance tradeoffs, *J. Oper. Manag.*, 24, no. 5, pp. 542–62, 2006.

[11] Clark K B and Fujimoto T, Lead time in automobile product development explaining the Japanese advantage, *J. Eng. Technol. Manag.*, 6, no. 1, pp. 25–58, 1989.

[12] Langerak F and Hultink E J, The effect of new product development acceleration approaches on development speed: A case study, *J. Eng. Technol. Manag. - JET-M*, 25, no. 3, pp. 157–67, 2008.

[13] Ellison D J, Clark K B, Takahiro F and Young-suk H, Product Development Performance in the Auto Industry: 1990s Update, *International Motor Vehicle Program*, Annual Spo. 1995.

[14] Cohen M A and Eliashberg J, New Product Development: The Performance and Time-to-Market Tradeoff, *Manage. Sci.*, 42, no. 2, pp. 173–86, 1996.

[15] Griffin A, "PDMA Research on New Product Development Practices: Updating Trends and Benchmarking Best Practices," *J. Prod. Innov. Manag*, 14, no. 6, pp. 429–58, 1997.

[16] Page A L, Assessing new product development practices and performance: Establishing crucial norms, *J. Prod. Innov. Manag.*, 10, no. 4, pp. 273–90, 1993.

[17] Wheelwright S C and Clark K B, Competing through Development Capability in Manufacturing-Based Organization, *Business Horizons*, 35, no. 4, pp. 29-43, 1992.

[18] Available from *http://www.maa.org.my/pdf/Market_Review_2017.pdf,* Accessed date 20th Dec, 2018.

[19] Available from *http://www.astroawani.com/berita-bisnes/proton-saga-paling-laris-pada-2017-165000*. Accessed date. 20th Dec 2018.

[20] Available from *https://www.motortrader.com.my/news/the-bestsellers-of-2008/,* Accessed date 04th Feb 2019.

Acknowledgments

This research was conducted with the support from the Perusahaan Otomobil Nasional Sdn. Bhd. (PROTON), and sponsored by the Universiti Teknikal Malaysia Melaka (UTeM).

IOP Conf. Series: Materials Science and Engineering **627** (2019) 012002 doi:10.1088/1757-899X/627/1/012002

An approach to determine condition of boring tool using acoustic & vibration signals

Pranjali S.Deole[1*], Shital S.Chiddarwar[2], Vishwas S.Deshpande[3]

1,3 Shri Ramdeobaba College of Engineering &Management,Nagpur-440013,India
2 Visvesvaraya National Institutes of Technology, Nagpur- 440012, India

* deoleps@rknec.edu

Abstract: The work presented in the paper focuses on an approach for automatic system to determine boring tool condition, using acoustic signatures. This acoustic signal based tool wear prediction system removes the need for manual inspection of the tool cutting edges and minimizes the time used on wear measurement. The tool condition prediction system proposed in this paper for boring operation essentially utilizes acoustic sensor i.e. microphone. In order to understand effects of vibrations generated during boring operation on tool, vibration signals were also recorded and processed to compare it with acoustic signals. The signals were acquired by data acquisition system in LABVIEW environment. The data thus obtained were further processed using Fourier transform and wavelet transform. Using this transforms,12 number of time domain, frequency domain and time frequency domain features were collected then with soft computing techniques which are learning from data strategy i.e.RBF and MLP, prediction of tool condition was done. The experimental results were analyzed with respect to various depth of cuts, feed rates and cutting speed for EN8,EN-9 and EN-31 steel. This approach is well suited for determining condition of boring tool using acoustic and vibration signals.

1. Introduction

The area of condition monitoring is gaining importance in the various machining operations because of the major concern in the manufacturing industries. Boring operations is one of the most essential machining operations in any manufacturing industry. Although the automated technologies has given great future as a cost effective solution, these systems as a whole could not be implemented full proof until certain prerequisites are met. The requirement of uninterrupted boring operation is one of the major prerequisite to achieve maximum efficiency. However, deteriorating process conditions such as tool wear resulting into bad geometric features for the part often results into the interruption of the machining operation causing increase in the downtime of the machine tool. Therefore, developing an effective tool wear monitoring has become one of the most important aspects in the automation of cutting/semi finishing processes. An effective condition monitoring system[1] must sense the condition of tool at appropriate time, allow for effective tool change strategy when tool becomes dull and maintain proper cutting conditions throughout the process. The monitoring of boring tool condition using acoustic signals is not much explored as a research area.The tool change policy in industry has resulted in many drawbacks like wastage of manufacturing resource if the tool is prematurely changed and hampering the quality of parts produced if the worn out tool is used for a longer time.This work aims to develop an approach to determine the condition of tool in boring operation by analyzing acoustic emissions. [2],[3] To understand the effects of vibrations generated during boring operation on tool, vibration signals were also recorded and processed to compare it with acoustic signals.

2. Approach

The proposed approach essentially considers boring as a machining operation to be done on CNC turning centre. In this work, indirect way of tool condition estimation is adopted hence acoustic as well as vibration signals are considered. The input parameters for machining are cutting speed (m/minute), feed (mm/revolution), depth of cut (mm), material hardness (HRC), tool slenderness ratio. The response parameters are surface finish and tool wear in terms of loss of weight. The responses are measured using industrial equipments and sensors.The details of the experiments are given in the next section. The signals acquired for analysis are acoustic emissions [4] and vibration, are intermediate responses from the system. To understand the relationship of the signals with output parameters, there is a need to establish statistical model between them. The result of the analysis and model building are discussed and the validation of the developed model using non linear system is also given in this paper.

3. Experimental Procedure

The experimental set up consists of a machine and sensors, the software systems, tools; raw material used.It also shows the machining parameters and the response parameters considered in the experiments. Figure 1 shows the schematic illustration of the experimental set up.

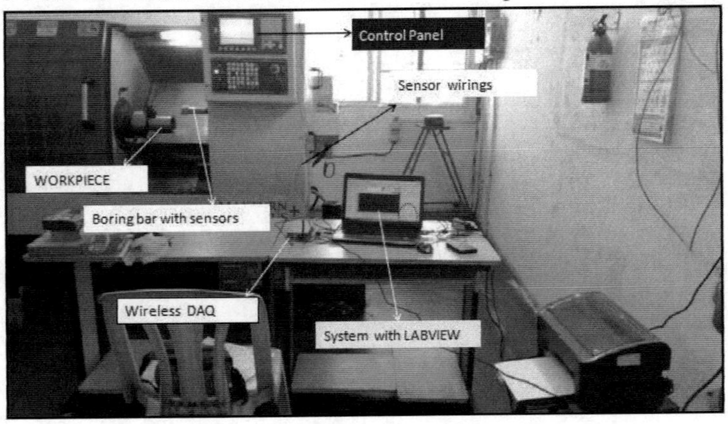

Figure 1. Experimental Set up

The experiments were carried out in the Laboratory environment.In this work a CNC Tuning centre (MTAB made,Model MAXTURN CNC Plus +) is utilized. The program of instruction for boring operation is developed for the workpiece.The machine has a Fanuc control system.The program is developed by using G codes and M codes given by the manufacturer. For experimentation purpose, a boring bar of 40 mm diameter. The boring operation is performed on the work pieces by using a CNMG 120408 Insert (CERATIZIT made). The material chosen for the experiments is EN 8,EN 9 and EN31). These are medium to high carbon steels and they are readily machinable in any conditions. The acoustic emissions are captured by means of an acoustic emission sensor and the vibration signals are captured by means of vibration sensor. A ¼inch free field microphone of GRASS made is utilized to capture the sound signals generated during machining of the workpiece, similarly a PCB made accelerometer is employed to capture the vibration signals during the cutting operations. For this study, NI Cdaq-9191 is used to acquire signals from the sensors. A virtual interface was created in the LABVIEW for acquiring the sound and vibration signals. The ultimate response parameters for the experimentation are tool wear and work piece surface roughness. Tool wear is calculated as a difference weights of the insert before machining and final weight of the insert i.e. the weight after machining. This difference is calculated for every experimental run. The weight of the insert was measured using the analytical balance. The surface roughness of the work pieces bored is measured by using a MITUTOYO made surface roughness tester, SJ201P.This method of measurement is a direct instrument measurement method. Three readings of the roughness average roughness parameter are taken and their average value is taken as roughness of the work piece.A sequence of experimental runs is generated using MINITAB software. [5] The independent variables [6] utilized in this study are: 3 cutting speeds (SP), 3 feed rates (FR), 3 depths of cut (DC), 3 material harnesses and 3 tool slenderness ratios and signals captured during machining. One insert was utilized for machining and the weight of the insert before and after machining was measured using an analytical balance. The measured values were used as the dependent variables in the statistical analysis, and as the output neuron in the ANN model.

Table 1. Scheme of experimentation.

Trial No.	Purpose	Input Parameters	Intermediate responses	Final Response
1	For data Collectio n	27*3=81 Runs(as per 3 steel materials)	Only acoustic signals	Surface Finish, Tool weight loss
2			Only vibration signals	
3		27 runs only for EN31	Acoustic and vibration signals both	
4			Only acoustic signals only for EN31	
5	For validation	Till tool worn out after 39 runs	Only acoustic emissions	Surface Finish, Tool weight loss
6		Till tool worn out after 31 runs	Only vibration signals	

For experimentation purpose, the conventional way of designing experiments was not considered. The main reason was the data required for prediction of tool condition during boring operation must be independent of design of experiments. Hence this scheme given in Table 1 was adopted. Initially, experiments were conducted and initial as well as final response variables were measured. For one trial that is 27 runs, the vibration channel of the data acquisition system was closed and only acoustic emission was recorded. In the 2nd trial, the acoustic sensor's channel was closed and only vibration data was recorded. During 3rd trial both the signals were read by the data acquisition system to collect the signals. The 4th trial was conducted only on EN31 with variable machining parameters. This was necessary to reproduce industrial scenario for boring operation. In all these four trials, the response variables were measured using the corresponding instruments. The data thus collected was further utilized for development of neural network model and building of a tool condition estimation system. For validation of the developed model, further experimentations were conducted. For this purpose, trial no. 5 and 6 were conducted till the tool insert completely worn out. These trial were conducted only on EN 31 with fixed values of input parameters. The reason behind this was to see the ability of the developed model to predict tool wear and remaining useful life.

4. Signal Processing

To develop a reliable tool condition monitoring system for predicting tool wear and useful remaining life of the tool, an approach of TCM is developed.The outline of the idea is presented in Figure 2. In this approach,the acoustic signals and vibration signals are captured from boring operation.These signals are then acquired through a wireless data acquisition system and are then transferred to computer for further signal processing.Based on the analysis of signals done,the decision making is done to predict the tool wear and remaining useful life of the tool.The response parameters surface finish and weight of the tool are measured and are related to the signals. The validation of the proposed model is done to ensure the viability of the system. So six important steps are developed for the proposed approach i.e.(a)Acquisition of signal data (b) Signal Processing (c)Feature Extraction (d) Feature selection and (e) decision making.

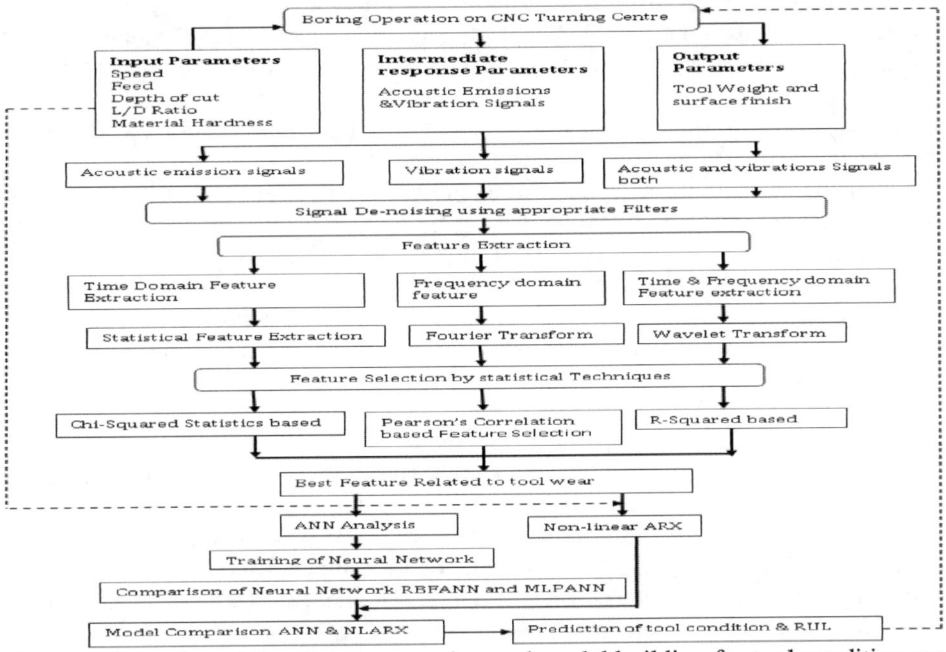

Figure 2. Procedure adapted for signal processing and model building for tool condition prediction

Signal preprocessing includes the amplification, de-nosing and conditioning of the acquired raw signal.The raw acoustic and vibration signal data is collected from the experimental runs performed using boring operation. The better the raw signal is de-noised, the more accurate the tool wear is predicted. he signals preprocessing includes signal de-noising using wavelet analysis. The denoising of the raw signal data is done by using wavelet analysis.

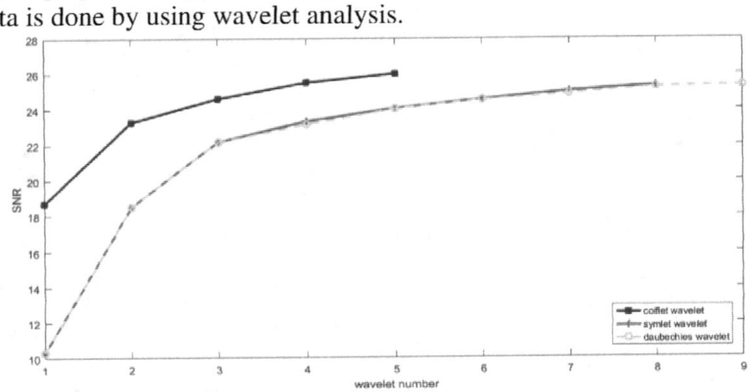

Figure3. Comparison of acoustic signal de-noising performance based on different wavelets

The comparison of acoustic signal denosing performance based on different wavelets i.e. Coifflet,Symlet and Daubechis wavelet is shown in the figure 34.The comparison of vibration signals denoising performance can also be obtained in the similar way.In order to understand the principles of monitoring strategies proposed in this study it is necessary to discuss some of the mathematical details regarding the feature extraction [7] and tool life detection strategies. Furthermore, a self organizing neural network is used for decision making regarding the tool wear and tool remaining useful life. The basic mathematical principle behind each of this aspect in the monitoring strategy is discussed.

4.1. Feature Extraction

The preprocessed signal contains large volume of information that's needs to be broken into meaningful information. This information so extracted are the features of the signals and are generally processed in time domain to obtain its statistical characteristics, in frequency domain to analyze power of signal distribution over a range of frequencies and in time frequency domain to obtain spectrum of non-stationary signals

4.2. Feature Selection

The tool life predictions with all the signal features are not the best selection because the irrelevant and redundant features are able to negatively affect the performance of a condition monitoring system and also the prediction model. The volume of the features extracted is very large; therefore, in order to improve the performance of prediction model, the features should be optimized. A numbers of techniques are used for this purpose. Here the techniques such as Pearson's correlation-based Feature Selection (CFS) method, chi-squared statistics selection method, R squared (R2) statistics selection method are used to optimize the features.

4.3.Model for Tool Life Estimation and Remaining Useful Life

To model TCM for estimating tool wear and remaining useful life of the tool, there was a need to use an information processing paradigm such as Artificial Neural network.ANN is inspired by the biological nervous system. The modeling of tool wear, RUL and surface roughness prediction system using ANN was essential due to highly non-linear relationship between inputs and outputs, over determined nature of the system and large size of data to handle. So two types of the neural network algorithm were used. Multi-layer perceptron neural network (MLP)and Radial basis function (RBF) [10] is utilized to model the tool condition and remaining useful life of the tool in boring operation.

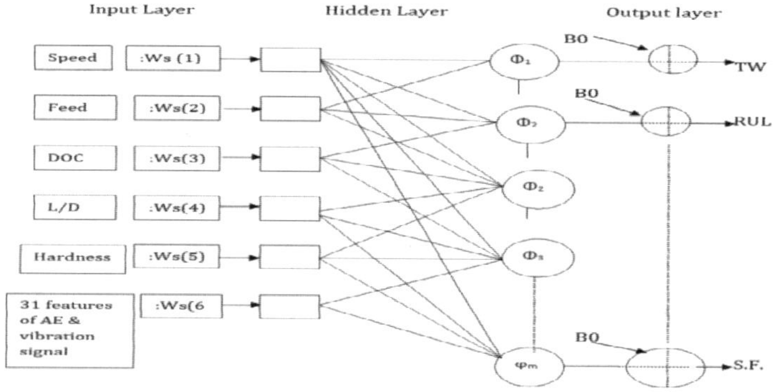

Figure 4 .RBF architecture used for tool condition prediction from acoustic and vibration signals

The experiments are conducted to collect the data of acoustic and vibration signals and corresponding output of weight loss of tool and roughness of the surface machined is recorded. It was found that the tool worn out at experiment no 39. This is found from the fact that there was a reduction of 0.05 gm of total weight of the cutting tool. The Figure 5 represents the tool wear and the tool worn out. It can be observed from the graph that tool worn out took place at 39th experimental run.

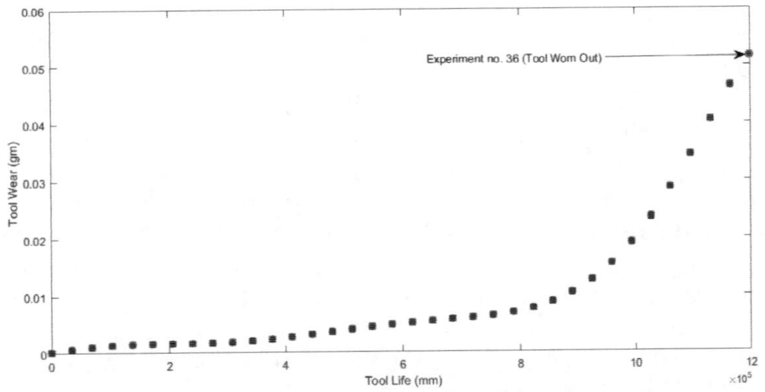

Figure 5. Tool Worn Out by the end of trial no 36.

As a same insert was used to carry out all the experimental runs, the insert started wearing out after carrying out experimental runs.As a consequence of which the remaining useful life of the cutting tool started reducing. The Figure 6 shows that when the tool wear reduces, the remaining useful life of the tool increases gradually. The red mark in the graph shows that there is a failure of tool at that point.

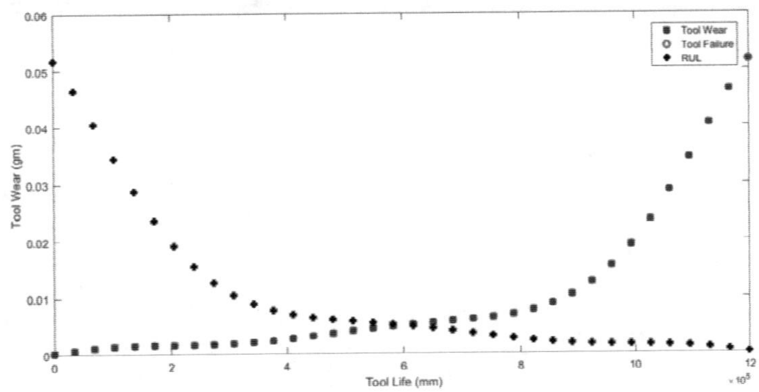

Figure 6. Variation of remaining useful life with. tool wear

The experiments are performed to acquire the acoustic signals and vibration signals both from the boring operation performed on CNC turning Centre. The signals data is acquired through a data acquisition system. The raw signals data is processed using advanced signal processing techniques. The acoustic and vibration signal data captured during boring operation is processed using appropriate ANN [1] signal processing technique.

When acoustic and vibration signals were captured during experiments with isolated machining parameters, following observations were made: The data collected from after feature extraction of the acoustic emission signals was used for prediction of tool wear. It is observed that the tool worn out at the end of Experiment no.39 which exactly matches with the experimental result. However, when same study was conducted with acquired vibration signals, the neural network designed for it predicted Tool worn out at the end of Experiment no. 31 which is 8 experiments prior to actual tool worn out. The correlation coefficient for first case was 0.9812and for second case 0.8036. This indicates that the vibration signal has some amount of noise remaining after de-noising. Thus acoustic signal becomes more reliable for accurate tool wear prediction under the considered scenario. This issue can be sorted out by further investigation of vibration signals [8]. For this, three components of the vibration signals (i.e. along x, y and z axis) needs to be captured separately.[9] This will help to remove unwanted signal and improve the reliability of vibration sensor. The results obtained from the signal processing techniques clearly give the indication of the applicability of the developed model.

5. Results and Conclusion

When acoustic and vibration signals were captured during experiments with isolated machining parameters, following observations were made: The data collected from after feature extraction of the acoustic emission signals was used for prediction of tool wear, it concluded that tool worn out at the end of Experiment no.39 which exactly matches with the experimental result. However, when same study was conducted with acquired vibration signals, the neural network designed for it predicted Tool worn out at the end of Experiment no. 31 which is 8 experiments prior to actual tool worn out. The correlation coefficient for first case was 0.9812and for second case 0.8037. This indicates that the vibration signal has some amount of noise remaining after de-noising. Thus acoustic signal becomes more reliable for accurate tool wear prediction under the considered scenario. This issue can be sorted out by further investigation of vibration signals. For this, three components of the vibration signals (i.e. along x, y and z axis) needs to be captured separately. This will help to remove unwanted signal and improve the reliability of vibration sensor.

Tool condition monitoring systems are finding vast applications in modern manufacturing industries due to their advantages in the production of quality parts and reducing waste of important manufacturing resources. As a very important part of the proposed TCM system, advanced signal processing technique was used to minimize the disadvantages of indirect monitoring method by proposing a wavelet packet transformation. The artificial neural network used in this work is easy to implement, highly accurate and reliable as compared to analytical methods. The developed systems of AE signal for TCM showed high-accuracy prediction capabilities for early detection of tool condition so the systems can be adopted as in-process tool condition monitoring systems to meet the needs of the cost-effective means to prevent defects and provide optimized tool usage strategy in the metal cutting industry.

6. References

[1] Francis, C., Uros, Z.,Real-Time Cutting Tool Condition Monitoring in Milling, Journal of Mechanical Engineering ,Volume 57,Issue 2, pp .142-150, 2011

[2] Li, X., Patri, K.V., Wavelet packet transforms of acoustic emission signals for tool wear monitoring. Journal for Manufacturing Science &Technology,Volume 1,Issue2,pp. 89-93,1999

[3] Mathew, M. T., Srinivasa Pai, P. and Rocha,L. A., An Effective Sensor for Tool Wear Monitoring in Face Milling : Acoustic Emission,Sadhana Volume 33, Issue 3, pp. 227–233, June 2008.

[4] Patra, K., Acoustic Emission based Tool Condition Monitoring System in Drilling, Proceedings of the World Congress on Engineering, London, U.K ,Volume 3, July 6 - 8, 2011.

[5] Botsaris, P.N., Tsanakas, J.A., Vogiatzi, M.E., Tool Wear Monitoring by Design of Experiments for Drilling, Proceedings of International Congress on condition monitoring and diagnostic engineering,Stavanger, Norway ,30 May -1 June, 2011.

[6] Panda,S.S., Chakraborty, D., Pal, S.K., Monitoring of drill flank wear in the time domain, Proceedings of The Int. Conf. On Artificial Intelligence And Applications,Innsbruck, Austria2005.

[7] Heinemann, R., Hinduja, S., A New Strategy for Tool Condition Monitoring of Small Diameter Twist Drills in Deep-Hole Drilling, International Journal of Machine Tools and Manufacture, Volume 52, pp 69–76, 2012.

[8] Pettersson, L., Vibration Analysis of a Boring Bar,Research Report, ISSN 1103-1581, Feb. 2002.

[9] Zhang, C., Yao, X., Zhang, J., Jin, H., Tool Condition Monitoring and Remaining Useful Life Prognostic Based on a Wireless Sensor in Dry Milling Operations, Sensors, Volume 16, Issue 6-795, pp 1-20, 2016.

[10] Xu C., Fan X., Luo W., Research of Tool Wear Based on Radical Basic Function Network,Proceedings of the International Multi Conference of Engineers and Computer Scientists 2009 Hong Kong,Volume I, March 18 - 20,pp.1-6, 2009.

IOP Conf. Series: Materials Science and Engineering **627** (2019) 012003 doi:10.1088/1757-899X/627/1/012003

Evaluation of brazed monolayer CBN grinding wheel on high-speed steel (HSS)

Bhaskar Pal[1], Muralidhar[2] and Nepal Adhikary[3]

[1]Professor, Presidency University, Bangalore, 560064, Karnataka, India
[2]Assistant Professor, Presidency University, Bangalore, 560064, Karnataka, India
[3]Director, R&D, Sim Yog Technology Private Limited, Bangalore, Karnataka, India
pbhaskar70@yahoo.com

Abstract. Shaping and forming of high-speed steel (HSS) tool with complicated geometry has been a challenge for the cutting tool industry. HSS grades generally display high hardness at high temperature and high wear resistance due to its constituent elements like a high amount of tungsten with the addition of chromium, vanadium and cobalt. HSS being used extensively in the manufacturing industry for machining difficult-cut-materials due to its high hardness and abrasion resistance. CBN grinding wheel is being used extensively for shaping the HSS. In this work, cBN brazed wheel with monolayer configuration is being used to grind HSS. Usually, reasonably uniform grit distribution, adequate spacing between the grits, 50-70% grit protrusion over the brazed surface and strong chemical bond strength are key features of brazed CBN wheel to perform satisfactorily. In the current investigation, the brazed CBN wheel showed limitations of its capability while grinding HSS in dry condition. The high chip load due to adverse grinding combinations, high attrition wear of grit due to its interaction with constituent elements like tungsten, chromium, vanadium and cobalt, fracturing and pull out of the grits are the limitation of the brazed wheel. The paper also concluded with facts to overcome limitations of brazed CBN wheel.

1. Introduction

There is an extreme demand for manufacturing HSS tool of complicated geometry to shape difficult-to-cut materials. The HSS has been one of the subsets of tool steels and is high-performance special steels, extensively used in cutting tool industry, has a high hardness at temperatures up to 500°C and high wear resistance. The alloying elements like tungsten, molybdenum, vanadium and chromium which can form strong carbides. The shaping and forming of the HSS tool always have been a challenge for the cutting tool industry. CBN is being used in resin, vitrified and metal bonded grinding wheel for grinding ferrous based alloy, titanium and nickel-based super-alloy in the various industrial applications due to its high hardness, chemically inactive towards ferrous alloys, unlike a diamond. There are ample of work done on a selection of grinding parameters, the effect on grit size for grinding steels [1][2]. In dry and wet grinding severe cracking on HSS observed but it was absent while grinding with cryo-cooling [3]. Work has been carried out in high-speed grinding, high efficiency deep grinding for alloy steel with conventional wheel, plated wheel and exhibited their performance [4]. The brazed monolayer CBN wheels have drawn considerable attention in the recent past due to its better grit distribution probability, 50-70% grit protrusion and strong chemical adhesion with base metal [5][6] An Ag-Cu base alloy, being activated with TiH2 powder, is the key alloy for brazing CBN grit with the steel base materials [7]. Induction brazing technique is being used for brazing cBN and residual stress has been estimated by finite element process [8]. It has established that vacuum brazing exhibited the best results. The process of uniform distribution of CBN grit has been well established on the periphery of the circular cylindrical job. This effective distribution of CBN grit on the brazed wheel has been the key parameters for its excellent performance [9][10]. The work has been carried out on titanium alloy, nickel-based superalloy [11][12] and different metallic materials [13] by monolayer CBN wheel.

All these work has been done very extensively with CBN brazed wheel but less work has been carried out to understand the performance of the brazed wheel on grit density. In this study, an attempt has been made to distribute the grit uniformly with specified grit density. This work has been carried out in said grit density and evaluating the performance of brazed CBN wheel on HSS work material in dry conditions.

2. Materials and Method

2.1 Preparation of sample

In the present study, a base alloy (72%Ag and 28%Cu) was activated by adding TiH_2 powder and form the paste. The wheel blank was placed in the milling indexing head for providing the rotational movement continuously. The paste (alloy) is being applied on the cylindrical surface of the wheel blank. The sieve was carefully attached on the x-y-z table of a travelling microscope and then placed on the periphery of the wheel. The CBN grits were poured regularly on the sieve. The grits were passed through the sieve and placed on the periphery of the wheel. This process is continued till wheel surface get uniformly distributed with grits [6].

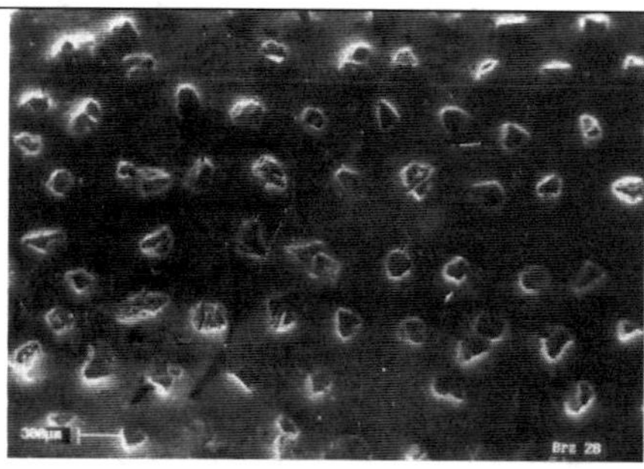

Fig.1. Distribution of cBN grit on the wheel

Fig.2 Grinding set up

Table I: Grinding Specifications

Brazed Grinding Wheel	Grits: Borazon 570 (GE Superabrasive, USA) B151
Wheel Diameter, d	15 mm
Work piece Material	High speed steel
Environment	Dry
Wheel-work piece contact width, w	3 mm
Length of cut per pass, L	80 mm
Speed of the spindle	Maximum speed 80000 rpm

Table II: Experimental Conditions

Wheel speed, v_s (m/s)	table speed, v_w, m/min	downfeed, d, µm	Remarks
Long duration test up to specific volume 100 mm^3/mm^2			
27.5	4	30	Adverse situation with high chip load
41	2	30	Favorable situation with less chip load

2.2 Brazing

Brazing process was conducted at 850°C temperature. The pressure in the furnace has kept around 10^{-5} torr to prevent oxidation during brazing. After completion of the brazing process, the brazed wheel was examined under Scanning Electron Microscope (SEM) [LEO 430i, England] as shown in Fig.1. It

IOP Conf. Series: Materials Science and Engineering **627** (2019) 012003 doi:10.1088/1757-899X/627/1/012003

observed that the process of seeding of the grits on the wheel has been done very effectively. Adequate spaces are provided between the grits for free accommodation chips to prevent loading during grinding operation [6]. The grit density has been calculated and it is found 8-9 grit per square mm of wheel surface. The assessment of cBN brazed wheel has been evaluated on this said grit density.

2.3. Grinding

The experiments were carried out in a surface grinding machine which was retrofitted with the high-speed spindle as shown in Fig.2. The spindle has been run at 20k, 35k, 45k and 52k to get the peripheral speed of 16 m/s, 27.5 m/s, 35 m/s and 41 m/s respectively. The maximum run out on the wheel was less than 5 μm. The high-speed steel (Length 80mmxWidth 3mmXDepth 20mm), was placed in a specially made fixture and mounted on a 3- D Kistler dynamometer [KISTLER 9257B, Switzerland] to measure the grinding forces as shown in Fig.2. The grinding specification and experimental combinations are indicated in Table I and Table II respectively.

3. Results and Discussion

3.1 Grinding forces

The magnitude of the tangential grinding force F_t and the normal grinding force F_n are recorded with the variation of the grinding process parameters, cutting speed, v_s, table speed, v_w and downfeed, d.

The magnitude of both F_t and F_n decreases with the increase in v_s for all combinations. This may be attributed to: i) it is evident from equivalent chip thickness ($t_e=v_w d/v_s$) that increase in v_s will lead to small chip thickness which leads to lesser chip load and ii) increase in v_s lead to high energy input which leads to plasticization of work material ahead of the grinding zone. But the forces increased with the increase in v_w and d for all combinations due to an increase in chip load.

The normal force, F_n, is always higher than the tangential force, F_t, in all conditions. The magnitude of the normal force, F_n, primarily depend on the hardness of the work material, grit tip geometry (sharpness), large negative rake angle of grit and resistance to penetration of the cutting points in the work surface. The tangential force, F_t, mainly depend on rubbing interaction with chip-grit, chip-bond interfaces and chip load. It is very much clear that the normal force component take major share to penetrate the work material and tangential force component follows the path to finish the work and formed chips.

The increase in table speed, v_w and downfeed, d, are desired for higher productivity. The productivity increases with the increase of v_w, and d which increases the chip load and consequently the grinding forces. It is evident that the grinding process has been discontinued in very adverse combinations like the low cutting speed, v_s, high table speed, v_w, and high downfeed, d, (v_s=16 m/s, v_w=3 m/min, d=40 μm and v_s=16 m/s, v_w=4 m/min, d=40 μm). It indicates that the performance of brazed cBN grinding wheel would be tested for long-duration test with specific grinding parameters to avoid premature wheel loading.

The long duration test has been conducted in two conditions:

i) Favorable condition (v_s=41 m/s, v_w=2 m/min, and d=30 μm) and t_e =0.024 μm.

ii) Adverse conditions (v_s=27.5 m/s, v_w=2 m/min, and d=30 μm) and t_e =0.036 μm.

In adverse condition, the equivalent chip thickness is 50% more than the favourable condition. It indicates that with said grit density the load per grit will be double. Effectiveness of brazed wheel depends on the sharpness of grit, cutting rather than rubbing & ploughing, grit density and wheel loading.

Fig.3(a) indicates the CBN wheel worked satisfactorily without any wheel loading in favourable conditions up to specific material removal of 100 mm^3/mm^2. As the less number of grit in action with the work material, the chip load per grit is much higher. Owing to high chip load and wider grit spacing, it has performed effectively. Fig.3(b) depicts the performance of adverse grinding conditions. It indicates that each grit has been subjected to a load which is twice that of favourable conditions.

Accordingly, grinding forces are relatively higher as compared with the favourable condition. Moreover, the HSS offered high resistance to grit penetration due to hot hardness and strong alloying elements like tungsten, molybdenum, vanadium and chromium carbides. The attrition wear rate at grit tip was excessive. Worn grit is incapable to penetrate into the work material and more rubbing, ploughing has happened than cutting resulted in high grinding forces in adverse condition. The grinding performance has been discontinued after 78 mm^3/mm^2 due to excessive grit wear as shown in Fig.3(b).

Favourable Conditions $v_s = 41\ m/s$ $v_w = 2\ m/min$, and $d = 30\ \mu m$.	Adverse Conditions $v_s = 27.5\ m/s$ $v_w = 2\ m/min$ $d = 30\ \mu m$.

Grinding forces (N) vs Specific Material removal (mm^3/mm^2)

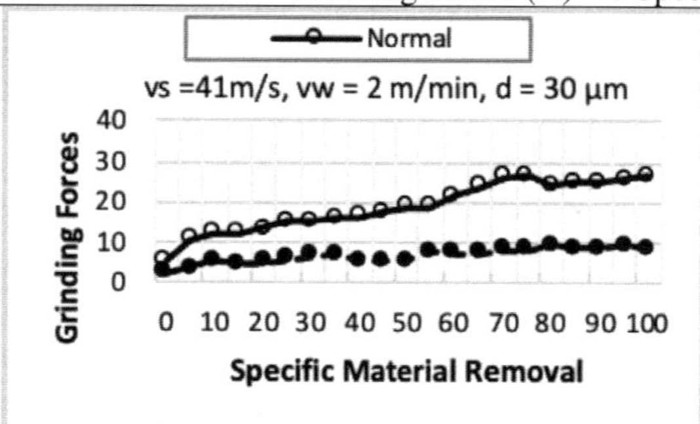

(a)

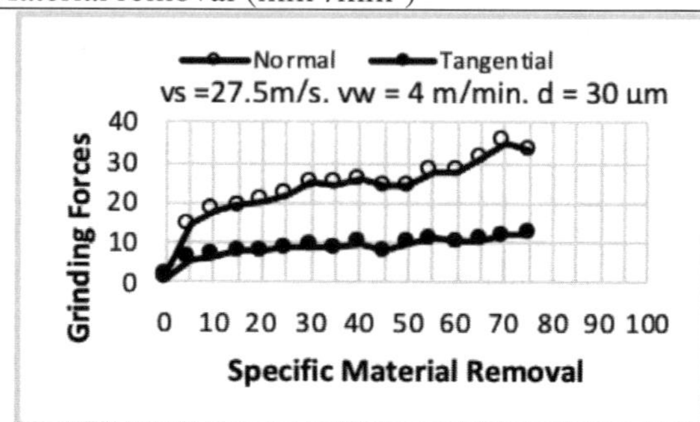

(b)

Wheel condition: specific material removal after 25 mm^3/mm^2

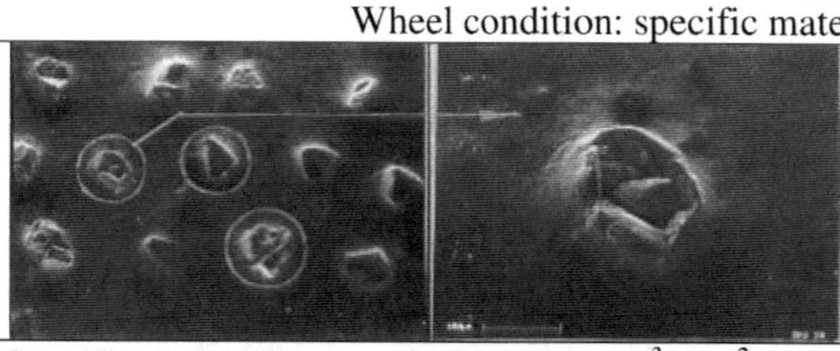

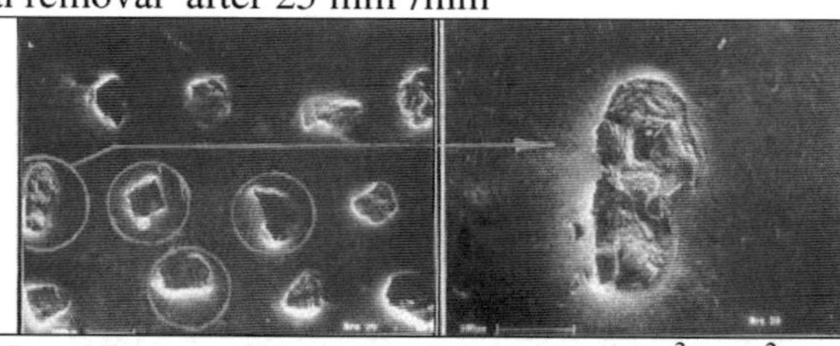

Specific material removal after 100 mm^3/mm^2

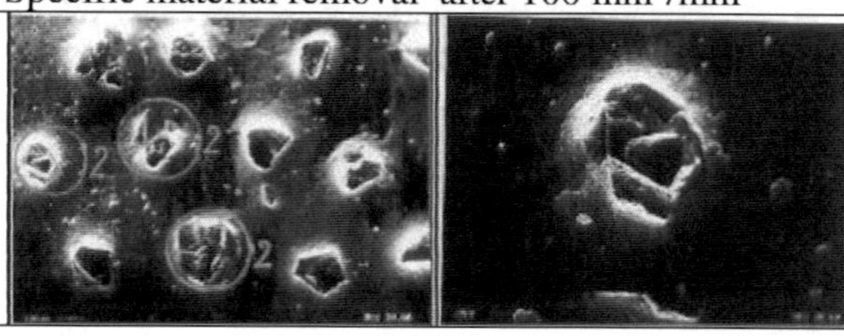

(c)

Specific material removal after 78 mm^3/mm^2

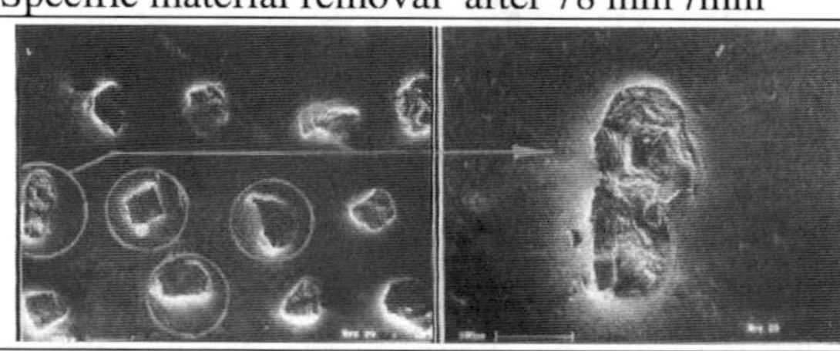

(d)

Fig.3. Condition of CBN grinding wheel with the progress of grinding HSS

3.2 Wheel topography

Fig.3(c) shows the active surface of the brazed CBN wheel after removing of specific material removal of 25 mm^3/mm^2 and 100 mm^3/mm^2 at the favourable grinding condition. No sign of wheel loading and grit pull out observed even up to the removal of 100 mm^3/mm^2. It discloses that the strong chemical bond has prevented dislodging of the grits from the bond but it reveals that the HSS caused a good level of fracture on the CBN grits. In Fig 3(d) it depicts that the grit wear very excessive. After 78 mm^3/mm^2 of material removal, the surface of the brazed wheel was remarkably deteriorated by wear of the grits. The high chip load and the hot hardness of the HSS are the possible reason for excessive attrition of the grits. The grinding performance has been aborted due to lack of cutting action.

4. Conclusions

a) The current process has been useful for achieving uniform distribution of grits and observed that 50-60 mesh size will realize the distribution of 8-9 grit per square per mm on the wheel surface.

b) The brazed wheel has been tested in two conditions i.e. i) Favorable condition (v_s=41 m/s, v_w=2 m/min, d=30 μm) and ii) Adverse condition (v_s=27.5 m/s, v_w=2 m/min, d=30 μm)

c) In favourable condition with said density, the grinding wheel has worked very effectively up to 100 mm^3/mm^2. Localized wheel loading, the interaction of work material with trapped chips, improper penetration of grit into work material and grit wear caused a sharp increase in the grinding forces (Fig.3(a)). The grits were flattened (Fig.3(a)) but still intact with the bond for continuing grinding process.

d) In adverse condition, the brazed wheel continued to grind the HSS up to 78 mm^3/mm^2 of specific material removal. Grinding process has been discontinued due to excessive wear of grit (Fig.3(d)).

e) In spite of large protrusion of the grit from bond and high forces acting on the grits, there was no significant grit pull out on the brazed wheel which indicates that the grits are strongly held with bond on its surface.

f) Higher grit density will reduce the chip load per grit resulting to enhance the overall service life of the brazed wheel in adverse condition by reducing attrition wear, fracturing and pull -out of the grits.

g) To derive maximum overall benefits of the brazed wheels, its configuration especially grit-spacing density need to be optimized through further intensive research.

5. References:

[1] S. Malkin, "Selection of Operating Parameters in Surface Grinding of Steels," *J. Eng. Ind.*, vol. 98, no. 1, p. 56, Feb. 1976.

[2] T. W. Hwang, C. J. Evans, and S. Malkin, "Size effect for specific energy in grinding of silicon nitride," *Wear*, vol. 225–229, pp. 862–867, Apr. 1999.

[3] S. Paul and A. B. Chattopadhyay, "The effect of cryogenic cooling on grinding forces," *Int. J. Mach. Tools Manuf.*, vol. 36, no. 1, pp. 63–72, Jan. 1996.

[4] D. J. Stephenson, T. Jin, and J. Corbett, "High Efficiency Deep Grinding of a Low Alloy Steel with Plated CBN Wheels," *CIRP Ann.*, vol. 51, no. 1, pp. 241–244, Jan. 2002.

[5] A. K. Chattopadhyay, L. Chollet, and H. E. Hintermann, "Improved monolayer CBN wheel for load free grinding," *Int. J. Mach. Tools Manuf.*, vol. 32, no. 4, pp. 571–581, Aug. 1992.

[6] B. Pal, "Development and Performance of Brazed Type Monolayer cBN Wheel." [Online]. Available: http://www.idr.iitkgp.ac.in/xmlui/handle/123456789/7701. [Accessed: 31-Jan-2019]

[7] W. F. Ding, J. H. Xu, J. B. Lu, Y. C. Fu, B. Xiao, and H. J. Xu, "Brazed CBN Grinding Wheel with Ag-Base Filler Alloy," *Mater. Sci. Forum*, vol. 471–472, pp. 11–15, Dec. 2004.

[8] W. XU, Y. ZHU, B. DU, and W. DING, "Residual stresses of polycrystalline CBN abrasive grits brazed with a high-frequency induction heating technique," *Chinese J. Aeronaut.*, Jul. 2018

[9] B. Pal, A. K. Chattopadhyay, and A. B. Chattopadhyay, "Development and performance evaluation of monolayer brazed cBN grinding wheel on bearing steel," *Int. J. Adv. Manuf. Technol.*, vol. 48, no. 9–12, 2010.

[10] B. Pal, A. K. Chattopadhyay, and A. B. Chattopadhyay, "Effect of chip morphology during grinding bearing steel using single layer brazed and galvanic bonded cubic boron nitride (cBN) wheel," *Mater. Manuf. Process.*, vol. 26, no. 8, 2011.

[11] U. Teicher, A. Ghosh, A. B. Chattopadhyay, and K. Künanz, "On the grindability of Titanium alloy by brazed type monolayered superabrasive grinding wheels," *Int. J. Mach. Tools Manuf.*, vol. 46, no. 6, pp. 620–622, May 2006.

[12] D. Wenfeng, X. Jiuhua, C. Zhenzhen, S. Honghua, and F. Yucan, "Grindability and Surface Integrity of Cast Nickel-based Superalloy in Creep Feed Grinding with Brazed CBN Abrasive Wheels," *Chinese J. Aeronaut.*, vol. 23, pp. 501–510, 2010.

[13] W. Ding *et al.*, "Review on monolayer CBN superabrasive wheels for grinding metallic materials," *Chinese J. Aeronaut.*, vol. 30, no. 1, pp. 109–134, Feb. 2017.

Mining and analyzing process similarity of product module for DPIPP based on PLM database

W P Peng[1], T Zhang[1,2], Q H Zhang[1] and S W Ni[1]

[1]School of Power and Mechanical Engineering, Wuhan University, 8 Donghunan Road, Wuchang District, Wuhan 430072, China

2017102080013@whu.edu.cn

Abstract. A distributed parametrized intelligent product platform (DPIPP) on basis of product lifecycle management database has been proposed to reveal the interrelationships between product attributes (i.e., structure, function or process) in recent years. Product modules in DPIPP are classified according to composition and structure of products. Components belonging to different product modules may have an identical or similar process route and could be manufacture with the same type of equipment. Mining and analysing the process similarity of product modules can effectively reduce production cost and improve production efficiency. In this paper, a novel product data mining approach with an improved Euclidean distance formula is proposed to research the product module process similarity, which aims to illustrate the effect of the structural change on the product process, then provide theoretical support for improving production patterns and reducing production costs to adapt to mass customization. The feasibility and effectiveness of the method are verified via the analysing of the process similarities among three different types of valves.

1. Introduction

In order to design and product rapidly, the similarity and classification method of product module have been widely studied. Several intelligent methods such as the genetic algorithm [1, 2, 3], the hybrid fuzzy clustering algorithm [4,5] and particle swarm optimization algorithm [6] has been used in this field. Product process route is the core of enterprise production management, so it is essential to carry out similarity analysis of product module process and merge similar process route, which could effectively reduce production coat. Lu et al [7, 8] constructed a new process model by taking advantage of an existing base model and process attribute information. Misaki et al. [9] established an effective method of searching the sheet metal parts based on bending process similarity. A similarity measure method of manufacturing process considering both unit similarity and sequence similarity on optimal matching of weighted bipartite graph was presented [10]. Li et al. [11] proposed a genetic algorithm, which could swiftly search for the optimal process plan for single manufacturing system as well as distributed manufacturing systems. Huang et al. [12] raised an algorithm for calculating process similarity in order to cluster process designs. Huang et al. [13] studied a novel numerical control machining process reuse approach by merging feature similarity assessment and data mining for computer-aided manufacturing models. Liu et al. [14] described a method of construction of product modules and analysis of process similarity. A rapid design technology of assembly process based on process similarity was studied [15].

In summary, most previous studies mainly concentrated on the similarity analysis of single attribute, but ignored the similarity analysis of multi-attribute. A distributed parametrized intelligent product platform (DPIPP) consisted of product modules, association rules and data dictionary has been proposed to satisfy the requirements of mass customization in recently years [16]. The product module in DPIPP is a generalized module which consists of parts or components from different products and their information (BOM, process card, process dictionary, function description etc.), and is also the carrier of function, structure and process of products [17]. The large amount of product information in the platform provides a basis for mining the similarity of multi-attributes.

In this paper, a novel product data mining approach with an improved Euclidean distance formula is proposed to analysis the similarity of multi-attributes among process and structure, aims at further optimizing DPIPP.

2. Description of process similarity of product module

2.1. Product module in DPIPP

The components A1-2, A2-2 from product A1, product A2 with related information constitute the product module C1 as shown in figure 1. So the product module in DPIPP can be expressed as follows:

$$g(M) = f(F_i, S_i, P_i) \tag{1}$$

where M is a certain generalized product module; $g(M)$ represents the main information of M; $\boldsymbol{F}_i$ is the ith function vector of M; $\boldsymbol{S}_i$ represents the ith structure vector of M; $\boldsymbol{P}_i$ is the ith process vector of M.

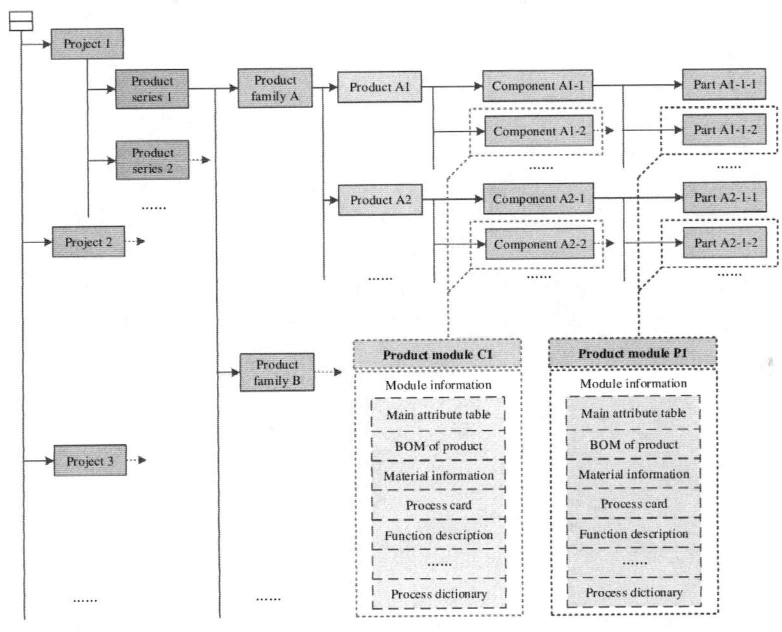

Figure 1. Product modules derived from the PLM database in DPIPP.

2.2. Process similarity of product module

Each product module in DPIPP contains the information of several parts or components. Therefore, there are several process routes in a product module. The process routes of the same product module are independent, but there may be some similarities among the process routes of different product modules. The process similarity of product modules is illustrated in figure 2, where M_1 is product module 1; M_2 is product module 2; The vector $\boldsymbol{P}_{i1}$ represents one of the process routes of the ith product module, in which the value is binary, 1 and 0 represent the process route contains some process technic or not. $\boldsymbol{P}_{11}$**&**$\boldsymbol{P}_{22}$ and $\boldsymbol{P}_{1j}$**&**$\boldsymbol{P}_{2j}$ are considered as the similar process routes through the comparison of all process information.

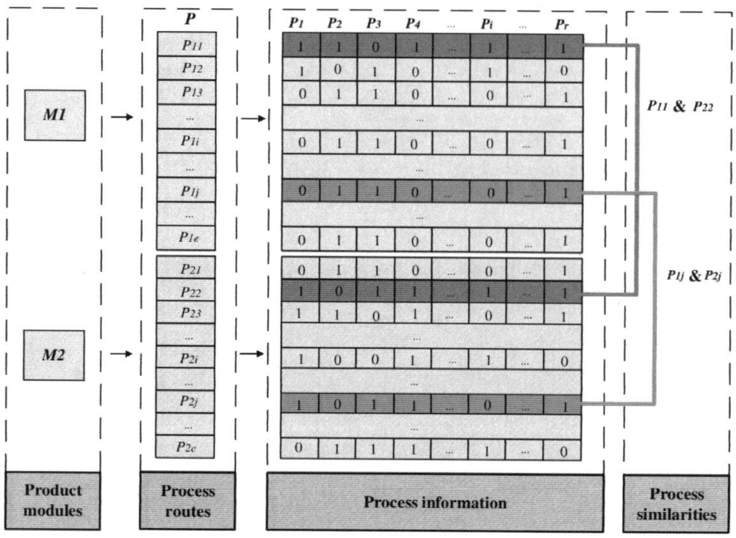

Figure 2. Process similarity of product module.

3. Process similarity analysis of product module

3.1. Process matrix constructing

Components in the same DPIPP product module have similar functions, thus the component structure becomes the primary factor affecting the process. To establish a structure matrix is the prerequisite for constructing a process matrix. For instance, M_x is a product module in DPIPP, the process parameters of M_x are affected by the following k structural parameters ($S_1, S_2,\ldots, S_k$, $k\in N^+$). The values of the k structural parameters can be represented as a vector. If there are m ($m\in N^+$) kinds of products with different structures in M_x, the number of structure vectors is m. The structural matrix of product module M_x is expressed by the m structure vectors as shown in equation (2):

$$M_S^x = \begin{bmatrix} S_{11}^x & S_{12}^x & \cdots & S_{1j}^x & \cdots & S_{1k}^x \\ S_{21}^x & S_{22}^x & \cdots & S_{2j}^x & \cdots & S_{2k}^x \\ \vdots & \vdots & \ddots & \vdots & \ddots & \vdots \\ S_{i1}^x & S_{i2}^x & \cdots & S_{ij}^x & \cdots & S_{ik}^x \\ \vdots & \vdots & \ddots & \vdots & \ddots & \vdots \\ S_{m1}^x & S_{m2}^x & \cdots & S_{mj}^x & \cdots & S_{mk}^x \end{bmatrix} \tag{2}$$

In above equation, $\boldsymbol{M^x_S}$ is the structural matrix of product module M_x; $\boldsymbol{S^x_j}$ is a vector of the $\boldsymbol{j^{th}}$ structure parameter of the product module M_x ; S^x_{ij} is the value of the $\boldsymbol{j^{th}}$ structural parameter of the $\boldsymbol{i^{th}}$ kind of product in product module M_x .

A structural vector of $\boldsymbol{M^x_S}$ corresponds to a process route, but different structures may share the same process route. Hence, the process routes of the m kinds of products in M_x might be less than m. If there are u ($u\leqslant m$, $u\in N^+$) different process routes in M_x, the process matrix can be expressed as follows:

$$M_p^x = [P_1^x, P_x^x, \cdots, P_i^x, \cdots, P_u^x]^T \tag{3}$$

where $\boldsymbol{M^x_P}$ represents the process matrix of product module M_x; $\boldsymbol{P^x_i}$ represents the $\boldsymbol{i^{th}}$ process route of $\boldsymbol{M_x}$ ($i\in(1,2,3\cdots u)$, which is composed of several process technics.

However, the process information is usually composed of numbers and words, in order to make the process matrix analysable, all the process technics should be defined in the process dictionary. Part of the process dictionary is shown in table 1. In the matrix $\boldsymbol{M^x_P}$, 1 and 0 represent weather a certain process technic contained in some process route or not respectively.

Table 1. Part of the process dictionary

P_1	P_2	P_3	P_4	P_5	……	P_r
Casting	Forging	Rough-milling	Finish-milling	Rough-turn	……	Finish-grinding

$$
M_P^x = \begin{array}{c} \begin{array}{cccccc} P_1 & P_2 & \cdots & P_i & \cdots & P_r \end{array} \\ \begin{bmatrix} P^x_{11} & P^x_{12} & \cdots & P^x_{1j} & \cdots & P^x_{1r} \\ P^x_{21} & P^x_{22} & \cdots & P^x_{2j} & \cdots & P^x_{2r} \\ \vdots & \vdots & \ddots & \vdots & \ddots & \vdots \\ P^x_{i1} & P^x_{i2} & \cdots & P^x_{ij} & \cdots & P^x_{ir} \\ \vdots & \vdots & \ddots & \vdots & \ddots & \vdots \\ P^x_{u1} & P^x_{u2} & \cdots & P^x_{uj} & \cdots & P^x_{ur} \end{bmatrix} \end{array} \rightarrow M_P^x = \begin{array}{c} \begin{array}{cccccc} P_1 & P_2 & \cdots & P_i & \cdots & P_r \end{array} \\ \begin{bmatrix} 1 & 0 & \cdots & 1 & \cdots & 1 \\ 1 & 1 & \cdots & 1 & \cdots & 1 \\ \vdots & \vdots & \ddots & \vdots & \ddots & \vdots \\ 0 & 1 & \cdots & 0 & \cdots & 0 \\ \vdots & \vdots & \ddots & \vdots & \ddots & \vdots \\ 1 & 0 & \cdots & 1 & \cdots & 1 \end{bmatrix} \end{array} \tag{4}
$$

3.2. Process similarity mining

In this subsection, the process route similarity degrees are calculated by the clustering algorithm [17] through the following two steps:

Step 1. Utilize an improved Euclidean formula to calculate the distances among different process routes;

Step 2. Convert the distances calculated in step 1 into the similarity degrees.

Suppose there are u and v process routes in M_x and M_y respectively ($u \in N^+$, $v \in N^+$). The process matrices are as follows:

$$
M_P^x = \begin{array}{c} \begin{array}{cccccc} P_1 & P_2 & \cdots & P_j & \cdots & P_r \end{array} \\ \begin{bmatrix} P^x_{11} & P^x_{21} & \cdots & P^x_{1j} & \cdots & P^x_{1r} \\ P^x_{21} & P^x_{22} & \cdots & P^x_{2j} & \cdots & P^x_{2r} \\ \vdots & \vdots & \ddots & \vdots & \ddots & \vdots \\ P^x_{i1} & P^x_{i2} & \cdots & P^x_{ij} & \cdots & P^x_{ir} \\ \vdots & \vdots & \ddots & \vdots & \ddots & \vdots \\ P^x_{u1} & P^x_{u2} & \cdots & P^x_{uj} & \cdots & P^x_{ur} \end{bmatrix} \end{array} \qquad M_P^y = \begin{array}{c} \begin{array}{cccccc} P_1 & P_2 & \cdots & P_j & \cdots & P_r \end{array} \\ \begin{bmatrix} P^y_{11} & P^y_{21} & \cdots & P^y_{1j} & \cdots & P^y_{1r} \\ P^y_{21} & P^y_{22} & \cdots & P^y_{2j} & \cdots & P^y_{2r} \\ \vdots & \vdots & \ddots & \vdots & \ddots & \vdots \\ P^y_{i1} & P^y_{i2} & \cdots & P^y_{ij} & \cdots & P^y_{ir} \\ \vdots & \vdots & \ddots & \vdots & \ddots & \vdots \\ P^y_{v1} & P^y_{v2} & \cdots & P^y_{vj} & \cdots & P^y_{vr} \end{bmatrix} \end{array} \tag{5}
$$

where $\forall P^x_{ij}, P^y_{ij} \in \{0,1\}$. A coefficient r is introduced into the Euclidean distance formula, which is convenient for standardizing the results of the calculation is used to calculate distances between two process routes from different product module in equation (6).

$$
d_{ij} = d(P_i^x, P_j^y) = \left[\frac{1}{r} \sum\nolimits_{w=1}^{r} (P^x_{iw} - P^y_{jw})^2 \right]^{\frac{1}{2}} \tag{6}
$$

where r refers to the number of process technics in Table 1.

The standardized distance matrix $\boldsymbol{d(M_x,M_y)}$ according to product modules M_x and M_x is expressed as follows:

$$d(M_x, M_y) = \begin{bmatrix} d_{11} & d_{12} & \cdots & d_{1j} & \cdots & d_{1v} \\ d_{21} & d_{22} & \cdots & d_{2j} & \cdots & d_{2v} \\ \vdots & \vdots & \ddots & \vdots & \ddots & \vdots \\ d_{i1} & d_{i2} & \cdots & d_{ij} & \cdots & d_{iv} \\ \vdots & \vdots & \ddots & \vdots & \ddots & \vdots \\ d_{u1} & d_{u2} & \cdots & d_{uj} & \cdots & d_{uv} \end{bmatrix} \quad (7)$$

where d_{ij} represents the difference between the ith process route of product module M_x and the jth process route of product module M_y .

Then the similarity degree s_{ij} can be calculated in equation (8):

$$s_{ij} = 1 - d_{ij} \quad (8)$$

The similarity degree matrix of product modules M_x and M_y can be established as follow.

$$Sim(M_x, M_y) = \begin{bmatrix} s_{11} & s_{12} & \cdots & s_{1j} & \cdots & s_{1v} \\ s_{21} & s_{22} & \cdots & s_{2j} & \cdots & s_{2v} \\ \vdots & \vdots & \ddots & \vdots & \ddots & \vdots \\ s_{i1} & s_{i2} & \cdots & s_{ij} & \cdots & s_{iv} \\ \vdots & \vdots & \ddots & \vdots & \ddots & \vdots \\ s_{u1} & s_{u2} & \cdots & s_{uj} & \cdots & s_{uv} \end{bmatrix} \quad (9)$$

The complex relationships of the structural matrices, the process matrices and the similarity degree matrix of product module M_x and M_y are clearly presented in figure 3.

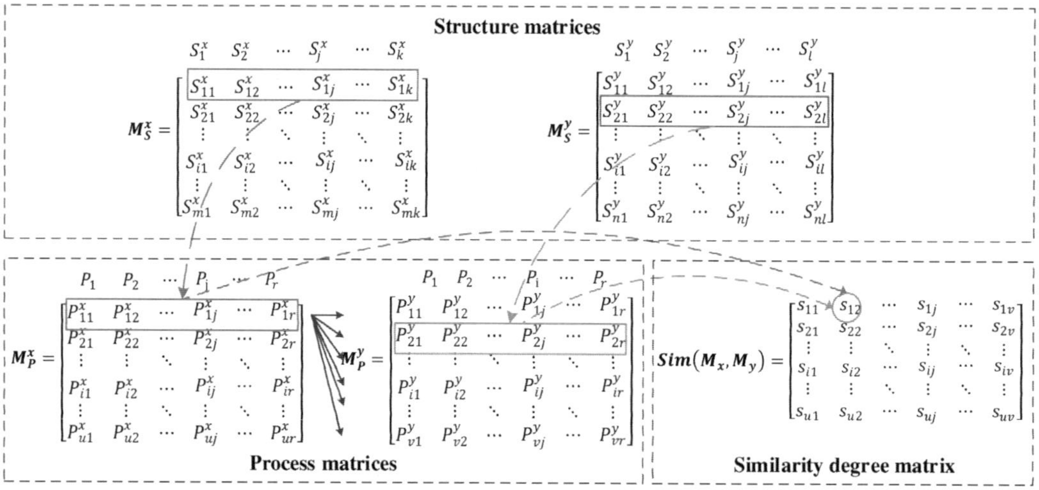

Figure 3. The mutual relationships of structural, process and similarity degree matrix.

Take the element s_{12} in the similarity degree matrix for instance, s_{12} represents the similarity of the 1st process route in product module M_x and the 2nd process route in product module M_y. The similarity degree matrix connects the structure matrices and the process matrices. Process similarity is an important design resource, plays a significant role in the production line optimization and rapid design. The analysis results of the process similarity are screened according to a certain similarity threshold that is set with the enterprise requirements, stored into a specific table in DPIPP.

4. Case

To verify the effectiveness of the process similarity analysis, an example of process similarity analysis of the product modules from a value company is presented. The typical structure of a valve is as follow.

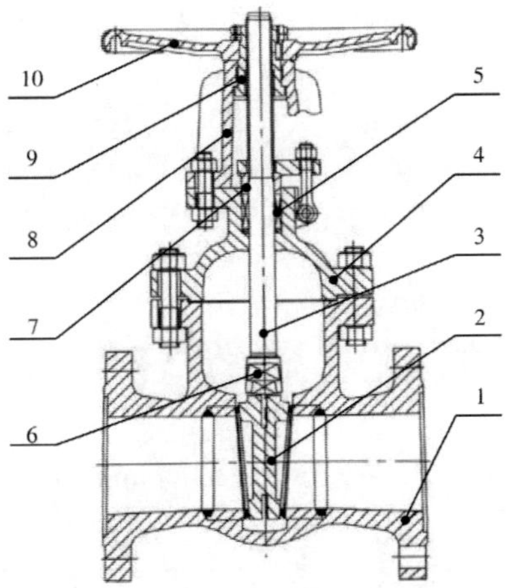

Figure. 4 The structure and components of a gate valve.
1.Valve Body; 2. Disc; 3. Stem; 4. Bonnet; 5. Seal ring; 6. Seat ring;
7.Stuffing box; 8. Bracket; 9. Stem Nut; 10. Hand Wheel.

4.1. Product module screening

The cavity of a valve is mainly composed of valve body and valve deck. The valve body is a vital part of a valve, bears the connection, sealing and supporting functions. In this section, the process similarities between three valve body modules are studied to illustrate the analysis process.

The valve body modules of gate valve, cut-off valve and ball valve are analyzed, which are expressed with M_z, M_j and M_q respectively. The structures of these three valves and the structural parameters which can influence the process of valve body are all shown in figure 5.

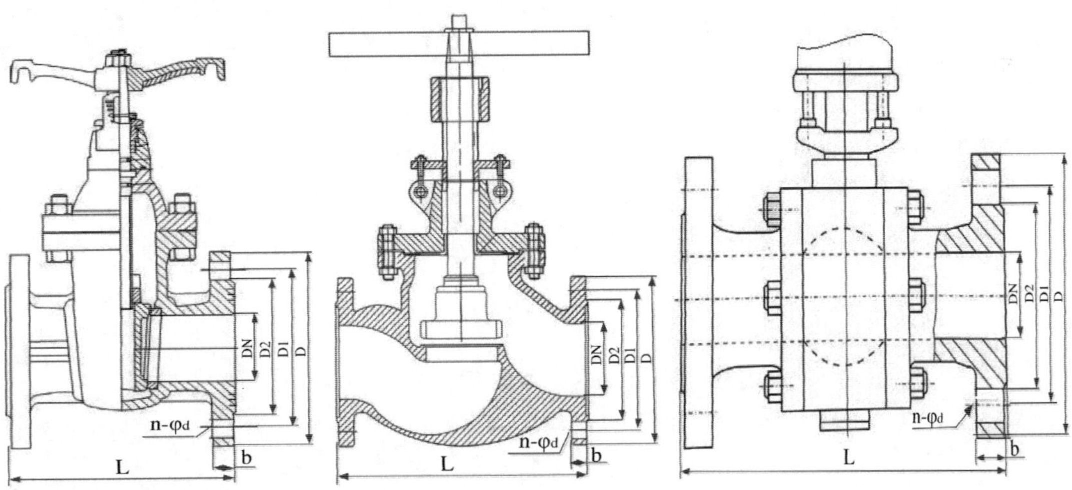

(a) Gate valve (b) Cut-off valve (c) Ball valve

Figure 5. The structures of the three valve bodies.

4.2. Process matrix constructing

The structural parameters of the three types of valves are extracted from the PLM database. These parameters will be stored in a unified format. Part of the structural parameters of the valve body for the gate valve is shown in table 2.

Table 2. Parts of the structural parameters of the gate valve's body.

No.	DN	L	D	D1	D2	b	n-φ_d
Z-01	50	250	165	125	99	20	4-18
Z-02	50	250	165	125	99	20	4-18
Z-03	65	265	185	145	118	20	4-18
Z-04	80	280	200	160	132	20	8-18
Z-05	100	300	220	180	156	22	8-18
Z-06	125	325	250	210	184	22	8-18
Z-07	150	350	285	240	211	24	8-22
……	……	……	……	……	……	……	……

Table 3. Process dictionary of the valve body module.

P_1	P_2	P_3	P_4	P_5	P_6
Bead weld	Turning	Chamfering	Rough-milling	Finish-milling	drilling
P_7	P_8	P_9	P_{10}	P_{11}	P_{12}
Rough-boring	Finish-boring	Reaming	Tapping	Ablation	Welding

The processes of the three valve body modules are encoded according to the process dictionary. Parts of the gate valve's process encodings are listed in table 4. There are some repeated process encodings as shows in table 4, such as the Z-03 and the Z-04. These repeated encodings will be merged in similarity analysis.

Table 4. Parts of the gate valve's process encodings.

No.	P_1	P_2	P_3	P_4	P_5	P_6	P_7	P_8	P_9	P_{10}	P_{11}	P_{12}
Z-01	0	1	1	0	1	0	0	1	1	1	1	0
Z-03	0	1	1	0	1	1	1	0	1	1	1	0
Z-04	0	1	1	0	1	1	1	0	1	1	1	0
Z-05	1	1	1	1	1	1	1	0	1	1	1	0
Z-06	0	1	1	1	1	1	1	0	1	1	1	0
Z-07	1	1	1	1	1	1	1	1	1	1	1	1
……	……	……	……	……	……	……	……	……	……	……	……	……

The process matrix is generated by the rest of process encodings. The process matrix of the gate valve can be expressed as follows.

$$M_P^Z = \begin{bmatrix} 0 & 1 & 1 & 0 & 0 & 0 & 0 & 0 & 1 & 1 & 1 & 0 \\ 0 & 1 & 1 & 0 & 0 & 1 & 0 & 0 & 0 & 1 & 1 & 0 \\ 0 & 1 & 1 & 0 & 1 & 1 & 0 & 0 & 0 & 1 & 1 & 0 \\ 0 & 1 & 1 & 0 & 1 & 0 & 1 & 1 & 0 & 1 & 1 & 0 \\ 0 & 1 & 1 & 0 & 1 & 0 & 1 & 1 & 0 & 1 & 1 & 1 \\ 0 & 1 & 1 & 1 & 1 & 0 & 1 & 1 & 0 & 1 & 1 & 0 \\ 1 & 1 & 1 & 1 & 1 & 0 & 1 & 1 & 0 & 1 & 1 & 0 \end{bmatrix} \tag{10}$$

4.1. Process similarity mining

The cavity of a valve is mainly composed of valve body and valve deck. The valve body is a vital part of a valve, bears the Connection, sealing and supporting functions. In this subsection, the process similarities between three valve body modules are studied to illustrate the analysis process. Following the construction of the three valves process matrices, the distances between these process matrices are calculated according to equation (6). Then these distances are transformed into the similarity degrees according to equation (8). The similarity degree matrices of the three product modules are as follows:

$$Sim(M_x, M_y) = \begin{bmatrix} 1.00 & 0.59 & 0.59 & 0.50 & 0.50 & 0.42 & 0.35 & 0.35 & 0.29 & 0.29 \\ 0.59 & 0.42 & 1.00 & 0.71 & 0.71 & 0.59 & 0.35 & 0.35 & 0.29 & 0.29 \\ 0.50 & 0.35 & 0.71 & 0.59 & 1.00 & 0.71 & 0.42 & 0.42 & 0.35 & 0.35 \\ 0.42 & 0.29 & 0.42 & 0.35 & 0.50 & 0.42 & 0.71 & 0.71 & 0.59 & 0.59 \\ 0.35 & 0.35 & 0.35 & 0.42 & 0.42 & 0.35 & 1.00 & 0.59 & 0.50 & 0.71 \\ 0.35 & 0.23 & 0.35 & 0.29 & 0.42 & 0.35 & 0.59 & 1.00 & 0.71 & 0.71 \\ 0.29 & 0.29 & 0.29 & 0.23 & 0.35 & 0.42 & 0.50 & 0.71 & 1.00 & 0.59 \end{bmatrix} \quad (11)$$

$$Sim(M_z, M_q) = \begin{bmatrix} 1.00 & 0.59 & 0.50 & 0.42 & 0.29 & 0.23 \\ 0.59 & 1.00 & 0.71 & 0.59 & 0.29 & 0.23 \\ 0.50 & 0.71 & 1.00 & 0.71 & 0.35 & 0.29 \\ 0.42 & 0.42 & 0.50 & 0.42 & 0.59 & 0.50 \\ 0.35 & 0.35 & 0.42 & 0.35 & 0.71 & 0.59 \\ 0.35 & 0.35 & 0.42 & 0.50 & 0.71 & 0.59 \\ 0.29 & 0.29 & 0.35 & 0.42 & 0.59 & 0.71 \end{bmatrix} \quad (12)$$

$$Sim(M_j, M_q) = \begin{bmatrix} 1.00 & 0.59 & 0.50 & 0.42 & 0.29 & 0.23 \\ 0.59 & 0.42 & 0.35 & 0.29 & 0.29 & 0.35 \\ 0.59 & 1.00 & 0.71 & 0.59 & 0.29 & 0.23 \\ 0.50 & 0.71 & 0.59 & 0.50 & 0.35 & 0.29 \\ 0.50 & 0.71 & 1.00 & 0.71 & 0.35 & 0.29 \\ 0.42 & 0.59 & 0.71 & 0.59 & 0.29 & 0.35 \\ 0.35 & 0.35 & 0.42 & 0.35 & 0.71 & 0.59 \\ 0.35 & 0.35 & 0.42 & 0.50 & 0.71 & 0.59 \\ 0.29 & 0.29 & 0.35 & 0.42 & 0.59 & 0.71 \\ 0.29 & 0.29 & 0.35 & 0.42 & 1.00 & 0.71 \end{bmatrix} \quad (13)$$

4.2. Process similarity storing

Finally, the similarity degrees of different product modules are all stored in DPIPP with the form of table 5, according to a certain similarity threshold μ, the product module with same process route.

Table 5. Parts of the process similarity information between gate valve body and cut-off body.

Similarity degrees	Two-dimension arrays		Corresponding relationships: Process codes (1,0, ……1)	Structural parameters $(S_1,S_2,……,S_{max})$	Product numbers
$S_{1,1}$=1.00	(z_1,j_1)	z_1	0,1,1,0,0,0,0,0,1,1,1,0	50,250,165,125,99,20,4-Φ18	x-1
				50,250,165,125,99,20,4-Φ18	x-5
				65,265,185,145,118,20,4-Φ18	x-7
				80,280,200,160,132,20,8-Φ18	x-12
		j_1	0,1,1,0,0,0,0,0,1,1,1,0	40,200,150,110,84,18, 4-Φ18	y-1
				50,230,165,125,99,20, 4-Φ18	y-8
				65,290,185,145,118,20, 4-Φ18	y-10
$S_{2,4}$=0.71	(z_2,j_4)	z_2	0,1,1,0,0,1,0,0,0,1,1,0	150,350,286,240,211,24,8-Φ22	x-3
				200,400,340,295,266,24,8-Φ22	x-6
				250,450,405,355,319,26,12-Φ26	x-10
				300,500,460,410,370,28,12-Φ26	x-15
		j_4	0,1,1,0,0,1,0,0,0,1,1,1	100,350,235,190,156,24,8-Φ22	y-6

				125,400,270,220,184,26,8-Φ26	y-15
				150,480,300,250,211,28,8-Φ26	y-18
……	……	……	……	……	……

5. Conclusions

In this paper, the product process similarity degree matrix related to structure is constructed on basis of the mining and analysing mutual relationship between process and structure of the product module in DPIPP. The product process information for string type is converted to binary by utilizing the process dictionary, which is convenient for compositing process matrix. The similarity degree between different process routes is calculated via an improved Euclidean distance formula, then the product module is clustered by a given threshold according to the practical requirement. The feasibility and the effectiveness of the approach has been verified though the valve case.

This research scientifically elaborates the influence of structure on process, and provide a theoretical support for further optimizing DPIPP through mining and analysing the product module process similarity. However, there are still many shortcomings that need to be improved in the following research, for instance, parameters of the product process and productive facilities should be considered because the same process may require different productive facilities due to the differences in process parameters. In addition, as the product data becomes more and more complex, intelligent product data mining methods ought to be apply to this field.

6. References

[1] Shi JY and Tao QB 2010 Research on a genetic algorithm for solving module partition problem *6th. Int. Conf. on Natural Computation* (Yantai: China/ New York IEEE) pp 4215-18

[2] Du H 2010 Application of GN Algorithm in Product Module Partition *International Journal of Digital Content Technology & its Applications* **6**(14)

[3] Xu ZW and Liu YX 2011 Study on Module Partition Method for CNC Lathe Design. *In Advances in Computer Science, Intelligent System and Environment* pp 749-753. (Berlin Heidelberg: Springer)

[4] Teng XY and Zhang TJ 2010 Module partition method based on hybrid fuzzy clustering. *Int. Conf. on Mechatronics and Automation* (Xi'an: China/ New York IEEE) pp 333-8

[5] Li H, Ji YJ, Gu XJ, et al 2012 Module partition process model and method of integrated service product. *Computers in Industry*; **63**(4) pp 298-308

[6] Shan Q, Chen Y and Yan GR 2008 Product Module Identification Based on PSO. *Int. Symp. on Information Science and Engineering* (Shanghai: China/ New York IEEE) pp 231-235

[7] Lu JD, Yao K and Gao FR 2009 Process similarity and developing new process models through migration. *AIChE journal* **55**(9) pp 2318-28.

[8] Lu JD and Gao FR 2008 Process modelling based on process similarity. *Industrial & Engineering Chemistry Research* **47**(6) pp 1967-74.

[9] Misaki D and Aomura S 2003 Searching sheet metal parts based on bending process similarity. *Proceedings of the Institution of Mechanical Engineers, Part B: Journal of Engineering Manufacture* **217**(3) pp 427-38.

[10] Zhang YJ, Liu JS, Wang L 2015 Product Manufacturing Process Similarity Measure Based on Attributed Graph Matching. *3rd Int. Conf. on Mechatronics, Robotics and Automation* (Shenzhen: China/ Paris: ATLANTIS) pp 1083-86

[11] Li L, Fuh J Y H and Zhang YF 2005 Application of genetic algorithm to computer-aided process planning in distributed manufacturing environments *Robotics and Computer-Integrated Manufacturing* **21**(6) pp 568-78

[12] Huang K, Zhou ZT and Han YB 2004 An algorithm for calculating process similarity to cluster open source process designs. *Grid and Cooperative Computing-GCC 2004 Workshops* (Wuhan: China/ Berlin Springer) pp 107-14

[13] Huang R, Zhang SS, Bai XL et al 2015 An effective numerical control machining process reuse approach by merging feature similarity assessment and data mining for computer-aided

manufacturing models. *Proceedings of the Institution of Mechanical Engineers, Part B: Journal of Engineering Manufacture* **229**(7) pp 1229-42
[14] Liu W, Liu ZY and Tan JR 2010Construction of Product Modules Based on Process Similarity. *Journal of Computer-Aided Design & Computer Graphics* **22**(10) pp 1647-54.
[15] Song HW, Zuo DW, Xue SL et al 2009 Research on rapid design technology of assembly process based on process similarity. *China Manufacturing Information: Academic Edition* **38**(11) pp 29-31.
[16] Huang YT 2015 Research on distributed parameterized intelligent product platform and its database realization (Thesis: Wuhan University)
[17] Cai RC, Zhang ZJ and Tung A K H 2014 A general framework of hierarchical clustering and its applications *Information Sciences* **272** pp 29-48.

Acknowledgments

This study was supported by National Natural Science Foundation (No.51275362).

IOP Conf. Series: Materials Science and Engineering **627** (2019) 012005 doi:10.1088/1757-899X/627/1/012005

Parametric optimization in Rotary Ultrasonic Drilling of BK7 through RSM based PSO

Hari Singh and Vikas Kumar
Department of Mechanical Engineering, NIT Kurukshetra, India

hsinghfme@nitkkr.ac.in

Abstract. This paper aims to establish the rotary ultrasonic machining as a pragmatic solution to tackle the machinability issues with brittle materials. In the present article, drilling variant of rotary ultrasonic has been used to machine the optical glass BK7 owing to its numerous applications in manufacturing sector. Experimental matrix has been designed by utilizing the response surface methodology after scrutinizing the machining variables (feed rate, spindle speed and ultrasonic power) through single variable at a time approach. Material removal rate (MRR) and surface roughness (SR) have been chosen as responses for assessing drilling performance. Thereafter ANOVA analysis was done at 95 % confidence level for figuring out the significant model terms (linear, interaction and square) that affect the responses of interest. Feed rate was found to have profound impact over the machining performance. Thereafter, desirability approach and particle swarm optimization (PSO) were used for yielding the right combination of input factors for concurrent optimization of responses. In comparison to desirability, particle swarm optimization (PSO) provided the instant optimal solutions to meet the industrial requirements. Further, confirmation experiments validated the accuracy of desirability and PSO to predict the optimal solution with little errors.

1. Introduction

With superior properties such as high hardness, high chemical resistance and wider transmission range with low thermal expansion co-efficient, BK-7 has numerous applications [1]. Conventional machining processes end up with many limitations like higher cutting forces, rapid tool wear, chipping and geometrical inaccuracies in machining BK-7 due to its high hardness and brittle nature, thus restricting its widespread applications. Even non-traditional machining processes including abrasive jet machining, chemical etching, electrochemical machining, laser beam machining, electric discharge machining, find it hard to become the pragmatic solution due to their inherent limitations like taper cutting, low material removal rate, surface defects, low aspect ratio, generation of heat affected zone and recast layer etc. Ultrasonic machining (USM) which is closest to rotary ultrasonic machining (RUM) in material removal mechanism suffers from low machining rate and higher cutting forces. So, need arises to explore the RUM to seek a fruitful solution to tackle all the problems mentioned above. RUM is an advanced machining process that involves direct contact of abrasive particles that are directly coated over the core drill. This arrangement results in hybridization of the conventional grinding and ultrasonic machining. This special arrangement also leads to achieve 6-10 times more material removal that cannot be matched by either conventional grinding or USM individually. Abrasive particles have light and intermittent contact with the workpiece surface and follow a spiral sinusoidal trajectory during the grinding variant of RUM [2].

1.1. Motivation for research work

Literature was extensively reviewed for identifying the following research gaps that led the authors to carry out the experimental investigation:

- Narrow range of ultrasonic power (20-40 %) in past research
- Scarcity of literature related to drilling of BK7 using RUM; rather researchers focused only on grinding variant of RUM for processing BK-7.
- Lack of use of techniques of design of experiments
- Almost no attempt on optimization of rotary ultrasonic drilling (RUD) performance using particle swarm optimization

To address the above-mentioned gaps, BK-7 has been drilled using the RUM with a broad range of ultrasonic power (30-80%). Further, response surface methodology (RSM) was coupled with desirability approach and particle swarm optimization for finding the right parametric setting of machining variables (ultrasonic power, spindle rotational speed and feed rate) for achieving concurrent optimization of responses of interest — material removal rate (MRR) and surface roughness (SR).

2. Materials and methods

2.1. Experimental set-up

Through holes of depth 5 mm were drilled through the square workpiece (50mm×50mm) using standard set-up of RUM (knee series 10, sonic mill), with a power rating of 1 kW. Enlarged view of machining zone and core drill are shown in Fig. 1. Core drills with 220 mesh sized diamond abrasives, 100% concentration and metal binder were used as constant tool variables for the machining. The outer diameter of core drill was 8 mm with thickness of 0.75 mm.

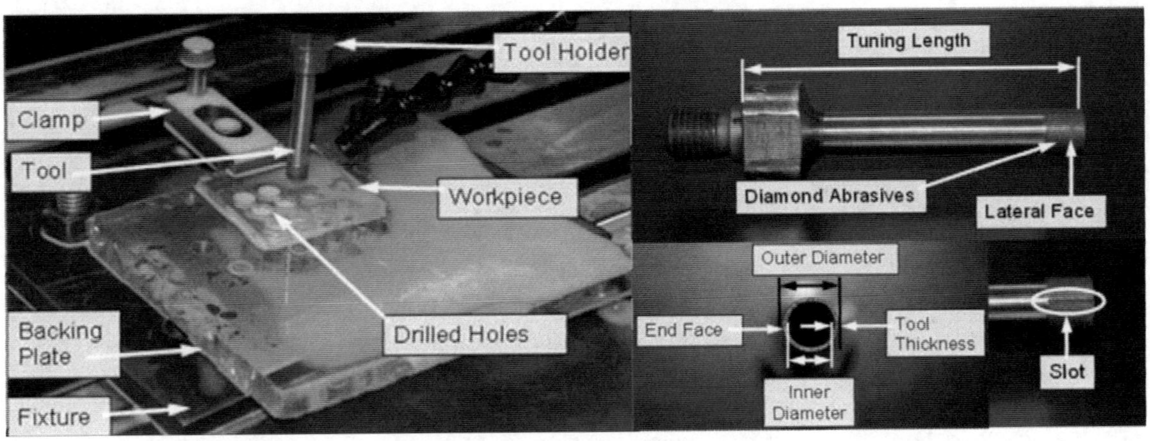

Figure 1. Machining zone and tool nomenclature

2.2. Experimental Procedure and data collection

The present experimental work involves the machining variables namely feed rate (FR), spindle speed (SS), and ultrasonic power (UP). Another machining variable "coolant pressure" was omitted from the final experimentation, as it did not show any remarkable influence over the responses of interest "MRR" and "SR" as revealed from screening/pilot experimentation. Design expert software (version 9) was used for statistical analysis and planning experimental matrix. Response surface methodology based central composite rotatable design (CCRD) was incorporated for framing the experiment settings as it best fits the second order response surfaces to estimate the regression coefficients accurately [3,4]. Table 1 gives the details of machining variables and their levels, which were finalized based on pilot experimentation and literature review.

Table 1. Machining variables and their levels

Machining Variables	Levels				
	L-1	L-2	L-3	L-4	L-5
SS- spindle speed (rpm)	2318	3000	4000	5000	5682
FR- Feed rate (mm/min)	0.095	0.30	0.60	0.90	1.105
UP- ultrasonic power (%)	30	40	55	70	80

3. Results and discussion

Experimentation was replicated twice and average of all the observations corresponding to each experimental setting is recorded in Table 2. Thereafter these observations were fed to design expert software to perform ANOVA analysis. The confidence level was kept 95% so as to apply the outcome of this research work on industrial scale. Consequently, model terms in Table 3 having P value less than or equal to 0.05 were significant and insignificant terms were removed by adopting the backward elimination method for improving the models. The results of ANOVA tests for both the responses are summarized in Table 3.

Table 2. Experimental observations for MRR and SR

Exp. No.	Spindle Speed (RPM)	Feed Rate (mm/min)	Ultrasonic Power (%)	MRR (mm^3/s)	SR (μm)
1	3000	0.3	40	0.1412	0.636
2	5000	0.3	40	0.1523	0.548
3	3000	0.9	40	0.2821	0.855
4	5000	0.9	40	0.3256	0.683
5	3000	0.3	70	0.165	0.398
6	5000	0.3	70	0.1936	0.385
7	3000	0.9	70	0.3926	0.808
8	5000	0.9	70	0.4869	0.779
9	2318	0.6	55	0.2093	0.714
10	5682	0.6	55	0.2906	0.559
11	4000	0.095	55	0.0891	0.368
12	4000	1.105	55	0.5003	0.914
13	4000	0.6	30	0.2011	0.735
14	4000	0.6	80	0.3563	0.524
15	4000	0.6	55	0.2283	0.593
16	4000	0.6	55	0.2426	0.590
17	4000	0.6	55	0.2381	0.577
18	4000	0.6	55	0.2463	0.583
19	4000	0.6	55	0.2253	0.550
20	4000	0.6	55	0.2548	0.609

3.1. Optimization through desirability and PSO

Desirability approach helps to attain the right parametric conditions when responses of interests are of conflicting nature as in present study where MRR increases with feed but at the cost of increase in SR [5]. It does so by transforming the Multi-objective function into a single objective function and then maximizing it. Although desirability gives the best possible optimal settings of input parameters to yield the better machining performance but finds it hard to give the optimal setting of one of the output parameters with desired values of other parameters. Such situation can be easily taken care of by PSO that is a stochastic intelligent technique [6,7].

Table 3. ANOVA for MRR and SR

Source	MRR			SR		
	SOS	P value	% contribution	SOS	P value	% contribution
Model	0.22	< 0.0001*	---	0.42	< 0.0001*	---
SS	0.00723	0.0001*	3.29	0.023	< 0.0001*	5.47
FR	0.17	< 0.0001*	77.27	0.32	< 0.0001*	76.19
UP	0.026	< 0.0001*	11.82	0.037	< 0.0001*	8.81
SS×FR	0.00120	0.0419*	0.55	Insignificant		
FR×UP	0.00534	0.0004*	2.43	0.025	< 0.0001*	5.95
SS×UP	Insignificant			0.00594	0.0078*	1.42
$(SS)^2$	Insignificant			0.00477	0.0143*	1.14
$(FR)^2$	0.004453	0.0009*	2.02	0.00564	0.0091*	1.34
$(UP)^2$	0.002038	0.0118*	0.93	0.00356	0.0289*	0.85
SS: spindle speed; FR: feed rate; UP: ultrasonic power; SOS: sum of squares; confidence level: 95 %						

For current investigation, first a PSO program was formulated using MATLAB (Version 8.0.1.604). MRR and SR models developed by RSM were fed to the program to look for optimal solutions. The optimization problem with objective function and all the constraints can be summarized as:

Maximize: MRR

$0.368 \leq SR\ (\mu m) \leq .914$; $2318 \leq$ Spindle speed (rpm) ≤ 5682 ;

$0.095 \leq$ Feed rate (mm/min) $\leq$ 1.105; $30 \leq$ Ultrasonic power (%) $\leq$ 80

Using these constraints and objective function, optimal conditions were predicted by PSO program for MRR by varying the SR from its minimum to maximum value. A graph as shown in Fig. 2 was plotted between MRR and SR for all the possible optimal solutions corresponding to a specific value of SR. Graph depicts the variation of the MRR with respect to SR. MRR was found to upsurge with increment in surface roughness, highlighting their conflicting nature. It happens due to increase in feed rate in the predicted solutions that reduced the machining time enhancing MRR but significantly deteriorating the surface finish. Optimal setting predicted by desirability was also validated using the PSO algorithm by putting constraint on SR as 0.531μm, which predicted the MRR value as 0.3483 mm^3/sec as shown in Fig. 2. Corresponding parametric condition predicted by PSO, logged in Table 4, was also validated experimentally at 95 % confidence level. This demonstrates the capability of PSO to predict almost similar value of MRR as predicted through RSM with the advantage of faster prediction as well.

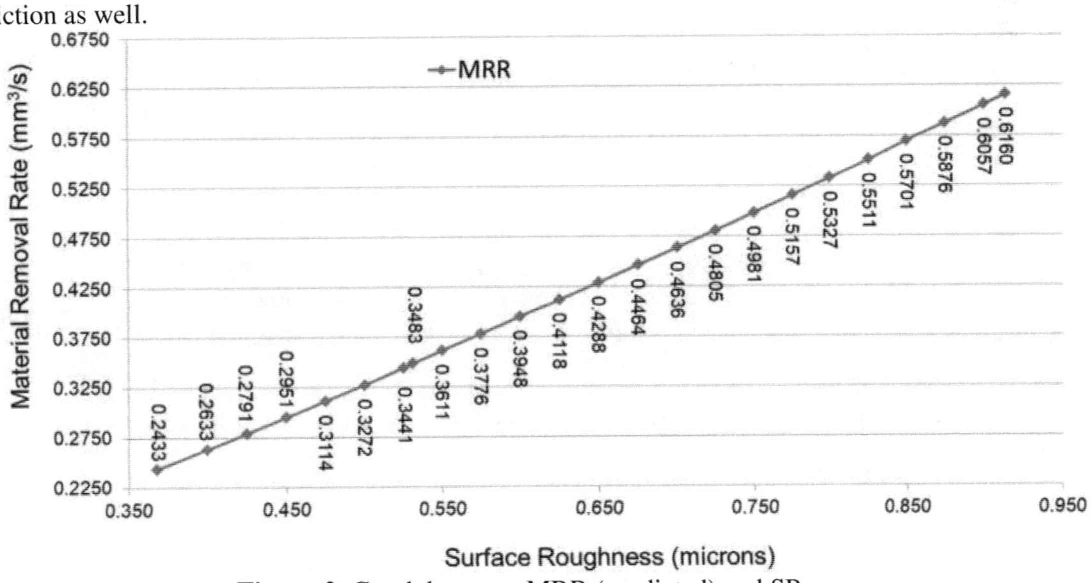

Figure 2. Graph between MRR (predicted) and SR

Table 4. Confirmatory results and optimized conditions for responses

Approach	Response(s)	Optimized conditions			Predicted	Experimental	%Error
		FR(mm/min)	SS (rpm)	UP (%)			
Desirability	MRR (mm^3/s)	0.57	4777	80	0.3485	0.3603	-3.38
	SR(μm)				0.531	0.546	-2.82
PSO	MRR (mm^3/s)	0.57	4831	80	0.3483	0.3641	-4.54
	SR(μm)				0.531	0.517	2.63

4. Conclusions

1. Selected machining variables (feed rate, spindle speed and ultrasonic power) have substantial impact on MRR at 95 % confidence level. The feed rate has maximum % contribution of 77.27 followed by ultrasonic power (11.82%) and spindle speed (3.29%) on MRR.
2. The feed rate had monotonic impact over surface roughness with maximum percentage contribution of 76.19 %. Ultrasonic power substantially affected the surface roughness with moderate contribution of 8.81 %.
3. The desirability approach provided optimal setting as spindle speed 4777 rpm, feed rate 0.57 mm/min and power 80 %.
4. PSO technique was equally viable for optimization as that of desirability with the advantage of faster prediction.
5. Current work was focused on machining variables only. In future work, some other input factors like abrasive size, type of abrasive bonding, tool diameter, tool thickness, tool profile can also be included to further improve the performance of RUM.

5. References

[1] Huu Loc P, Shiou F, Yu Z and Hsu W 2013 Investigation of Optimal Air-Driving Fluid Jet Polishing Parameters for the Surface Finish of N-BK7 Optical Glass J. Manuf. Sci. Eng. 135 1–7

[2] Wang J, Feng P, Zhang J, Zhang C and Pei Z 2016 Modeling the dependency of edge chipping size on the material properties and cutting force for rotary ultrasonic drilling of brittle materials Int. J. Mach. Tools Manuf. 101 18–27

[3] Chittaranjandas V 2016 Response surface Methodology and Desirability Approach to Optimize EDM Parameters Int. J. Hybrid Inf. Technol. 9 393–406

[4] Zahrani E G and Marasi A 2012 Modeling and optimization of laser bending parameters via response surface methodology Proc. Inst. Mech. Engrs Part C J Mech. Eng. Sci. 227 1577–84

[5] Akteke-Öztürk B, Weber G-W and Köksal G 2015 Desirability Functions in Multiresponse Optimization Communications in Computer and Information Science vol 499 (Springer International Publishing) pp 129–46

[6] Wen X L, Huang J C, Sheng D H and Wang F L 2010 Conicity and cylindricity error evaluation using particle swarm optimization Precis. Eng. Int. Soc. Precis. Eng. Nanotechnol. 34 338–44

[7] Chen G, Xiong B and Huang X 2011 Finding the optimal characteristic parameters for 3R pseudo-rigid-body model using an improved particle swarm optimizer Precis. Eng. 35 505–11

Acknowledgment

Authors are grateful to NIT Kurukshetra for providing the necessary financial support.

Research of the Green Product Design and Evaluation Methods based on Concurrent Engineering

Li Xiaona[1], Liu Xin[2]

1 School of Design Tianjin Commerce University, Tianjin, China
2 Department of Design Center of Tianjin Optical Electrical Group Co.,Ltd., Tianjin, China
E-mail: xnsmile@163.com

Abstract. Based on the concept of concurrent engineering and green design, the framework of green concurrent engineering structure was discussed. Then the key technologies and key factors were analyzed, and the evaluation indicators and design methods of green design are studied. So product development could be controlled from the whole process in order to improve the development efficiency and product quality, and reduce the pollution to protect the environment.

1. Introduction

With the technology development and consumer's updated demands, manufacture industry is facing more and more challenges. Even though the rapid development of manufacturing industry promotes human material progress a lot, its destruction on the ecological environment cannot be ignored. The excessive consumption of energy and resource, environmental degradation and strict environmental protection laws and regulations push the manufacturing industry transform from extensive development type to green design and green manufacturing. Research has shown that the cost in the stage of product development, including concept design, preliminary design and detailed design, accounts for only 18% of the total cost, but it will determine the product's value of 80% - 90% [1]. Similarly, the concept of green design must run throughout the whole life cycle, and the whole product development should be designed and managed, which is the important idea of concurrent engineering. Therefore, the concurrent green design method emerged. The design and evaluation method of green concurrent engineering will be mainly discussed then.

2. Green concurrent engineering

Concurrent Engineering, also known as Simultaneous Engineering(SE) , a systematic integration method to support product design and manufacturing process in parallel way, namely considering product design and manufacturing system at the same time, to ensure that the product structure in accordance with its function and the requirement of manufacturing process. It includes all the upstream and downstream of the product life cycle activities, that is, from market analysis, raw materials acquisition, product cost analysis, product processing, assembly, inspection, until the product after-sales service, maintenance, and finally retired, scrap recycling, etc [2-3].

Green design, also known as ecological design or environmental design, first appeared in an environmental pollution regulation of the United States in the 1970s. Green design is a design process centering on environmental resources, that is, during the whole life cycle of the product, priority should be given to the product's environmental attributes, disassembly, recyclability, etc., which as product design goals. While meeting the environmental objectives, ensure the physical objectives, basic performance, service life, quality and so on [4]. Common contents of green design include: green design material selection, design for disassembly, recycling design, design for manufacturing and assembly, productcost analysis, life cycle evaluation, etc [5]. Compared with conventional design, green design takes more consideration of energy consumption, resource utilization, environmental pollution, human health impact and ecological benefits in the whole life cycle of products.

It can be seen from the above that green design and concurrent engineering essentially have same goal. Therefore, the combination of the two to jointly complete product development and design will certainly improve efficiency and competitiveness. Therefore, green concurrent engineering is an engineering methodology aiming at the serial production mode of traditional products based on the whole process of product design, manufacturing and recycling. Although winners,R,I and others have come up with the concept, there is no clear definition yet [6]. It can be considered that concurrent engineering is a systematic approach to design products and related processes in an integrated and parallel manner. And green concurrent engineering firstly requires green product developers to consider all the factors from concept formation to product obsolescence in the entire product life cycle, including quality, cost, schedule, user requirements, environmental impact and recycling. And so in the early stage of product development, problems in the whole process of product development can be found timely, which shortens the product development cycle and enhances the competitiveness of enterprises in the market.

3. Research on key factors and methods of green concurrent engineering

3.1. green concurrent engineering framework

Green concurrent engineering is essentially an advanced design technology based on the organic combination of "green" and "concurrent", which fully reflects the overall advantages of greenization, parallelization and integration. From the perspective of product design, the development process can be divided into planning stage, conceptual design and detailed design, model making and prototype making, mass production and product recovery during the whole life cycle of products [7]. Combined with product life cycle and concurrent engineering, the green concurrent engineering, including product design, product structure design, manufacturing process design, manufacturing environment design, packaging design, use and maintenance design and recycling design, can be summarized as Figure 1.

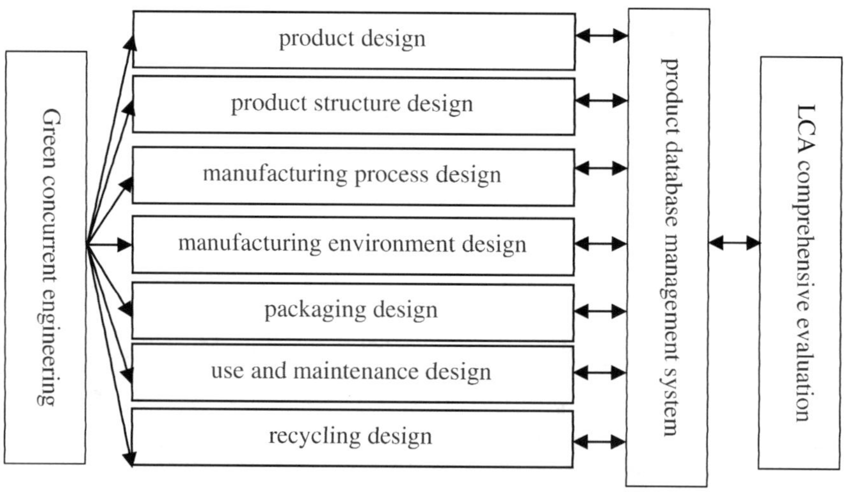

Figure 1. Green concurrent engineering framework.

The process of using concurrent green design products is divided into several stages. Each stage is divided into several stages, each of which has its own time period and central task, and some of the stages overlap with each other. The overlapping part indicates that the jobs in these stages are carried out in parallel. In general, the two or more adjacent phases can overlap with each other. Through simulation, evaluation, decision making and optimization design, the product can form a closed loop process from conceptual design to recycling treatment after the end of life, so as to meet the green requirements of the whole product life cycle.

3.2. Key factors study

3.2.1 Process alignment and goal alignment. Green concurrent engineering emphasizes that it should be oriented to the whole process or product object. Therefore, it emphasizes that designers should not only consider the design, but also consider the feasibility of various processes. That is, in the design process of products, the technology, manufacturability, productivity and maintainability of the design are mainly considered. People in the process department should also consider other processes. When designing a part, they should consider the cooperation with other parts. The time-sequenced engineering process is transformed into simultaneous consideration and concurrent operation, especially concurrent coordination and synchronous design at the initial stage of product design. Take into account the whole product life cycle (technical design/process design/manufacturing and testing/use and maintenance/recovery and treatment), and focus on environmental protection. So concurrent engineering looks at the whole process and the target, and considers both.

3.2.2 Database setup and system management. Green design data refers to the data generated and used in the design process. Green design knowledge refers to the rules needed to support green design decisions. Since green design involves the whole product life cycle, the data and knowledge needed for design is the fusion and integration required for each stage of the product life cycle. In order to meet the needs of green design, a corresponding database and knowledge base must be established to provide important data support and knowledge support for scheme design, structural design, material design, process design, packaging design, use design and recycling design of green products. The main contents of green design include: disassembly design database, recovery design database, operation design database, material selection database, green knowledge base, etc. In order to realize seamless integration of various design technologies, LCA analysis results are directly used to guide product assembly design, and product data management technology is applied to concurrent green design. Product data management (PDM) system can integrate all kinds of information, such as CAD/ CAPP/CAM/ CAE file, material list, product specification, nc program, production cost, etc., into a unified platform to provide information management of the whole life cycle of products. PDM is the foundation of the new technology such as product restructuring, concurrent engineering, collaborative product development, which can effectively make the definition, organization and management of related data, make the data consistent, sharing and innovation from product concept design, calculation and analysis, structure design, process, manufacturing, sales, maintenance services and product waste reprocessing of the whole process . It also and can track, conservation and management of large amounts of data required of product design, manufacturing and technical support, to control the processing and use of the product information, and optimize the product development.

3.2.3 Information integration and team collaboration. Due to the lack of functional interaction and the management of collaborative activity in traditional CAD and CAM systems, it is difficult to realize the interaction between functions and design process, and the concurrent effect is poor. To realize concurrent design, CAD personnel must control the design process while carrying out green design, so that CAPP and CAM personnel can participate in the green design of products, and guide designers' behavior by means of mutual information transmission, knowledge interaction or discussion. Through PDM platform, it can provide each design team with a flexible product data management tool and integrate all designers into a unified environment. This requires the full use of computer network and Internet technology, and the use of distributed database technology and object-oriented methods to design the system, so as to form an integrated and concurrent product design and development environment.

3.2.4 Green design evaluation. In order to realize green design, the environmental impact of the designed product in its life cycle should be firstly analyzed, and the evaluation index and evaluation methods should be determined. Secondly, the design should be improved according to the analysis results. The next chapter goes into details.

4. Research on evaluation index and method of green design

There are various types of products, and a scientific and comprehensive evaluation index system is the first problem to be solved in green comprehensive evaluation of design scheme. This paper takes mechanical products as an example. In such products' evaluation index system building research, Zeng XH [8] and other scholars believe that green packaging machine evaluation should include: the clean production evaluation, product flow evaluation, product use and maintenance evaluation, environmental burden evaluation, environmental value evaluation and the evaluation of the environmental impact; Huang DZ [9] built a comprehensive evaluation system for green design selection of engineering machinery, which includes resource attribute index, energy attribute index, economic index and environmental protection attribute index.

The design of index system should follow the comprehensive principle, scientific principle, feasibility principle and the combination of qualitative and quantitative principle. Based on the relevant literature and the analysis of the green system of mechanical products, this paper believes that the green design of mechanical products based on the concurrent engineering should be comprehensively evaluated from four aspects: resource attribute index, energy attribute index, economic index and environmental protection attribute index. Among them, the resource attribute index mainly considers the material resources and device resources used in each stage of the product life cycle. The index of energy attribute mainly includes two aspects, one is the utilization rate of energy, which reflects the problem of energy conservation. Second, the utilization rate of clean energy, which reflects the environmental protection problem. Economic index mainly reflect the extent to which products can be disassembled and reused after being scrapped, including the cost of product design, disassembly of products and utilization rate after being scrapped. Environmental attribute index mainly take into account the indicators related to the environment of products in the whole life cycle, such as environmental pollution, ecological damage and physical health. The related database of green design evaluation system includes: green design database, green process database, green assembly database and green recovery database. The structure of the index system is shown in Figure 2.

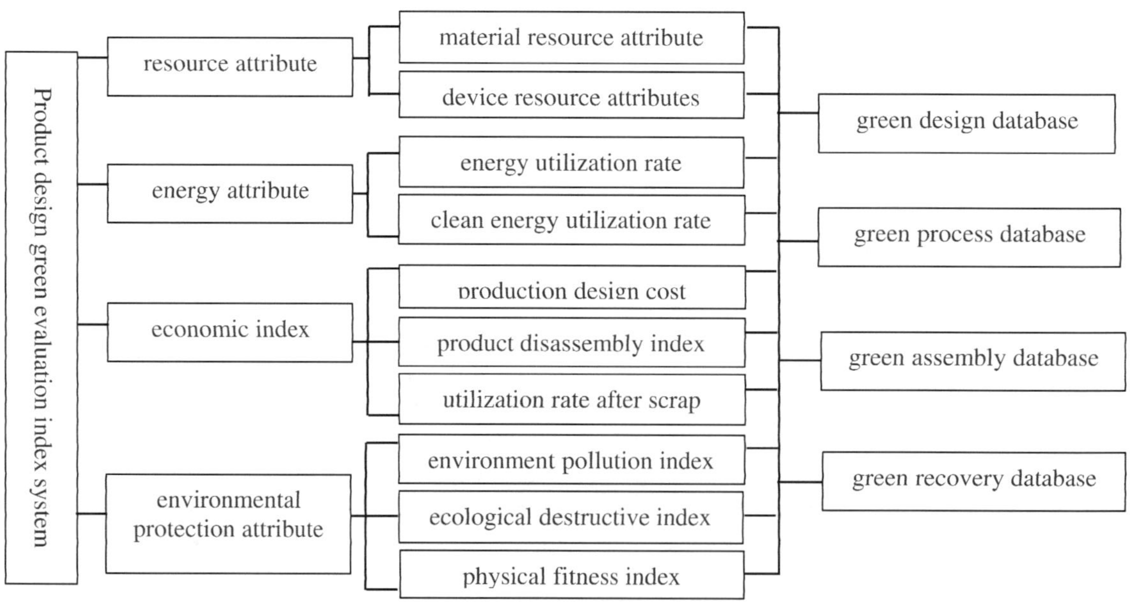

Figure 2. Product design green evaluation index system.

Green design evaluation system is designed, through the establishment of technology- economy-environment model, to provide support information for research and development of green products, to evaluate the new product's technical, economic and environmental coordination, to determine the green degree of new product design, to make comprehensive evaluation and comparison of design scheme selection, and to provide reference and basis for the design of green products. In terms of research on comprehensive evaluation methods of green design scheme of mechanical products, Zeng XH [8] proposed the green degree fuzzy comprehensive evaluation model and evaluated the packaging machinery products. Chen M [10] used AHP and fuzzy comprehensive evaluation methods to build a comprehensive evaluation model of product design scheme greenness.

Based on concurrent engineering, the feature model of product life cycle can be integrated into one parameter by using the multi-objective evaluation method. There are three objective functions, that is, the product recycling time (f_t) is the longest, the resource comprehensive utilization rate (f_r) is the highest, and the environment negative impact (f_e) is the lowest within the product life cycle. There are mainly 5 constraints, namely, product function (F), delivery time (T), quality (Q), cost (C) and service (S) reach the corresponding target value. The calculation method of product life cycle engineering characteristic model is as follows [11].

$$\begin{cases} \text{Max V} = \{f_t(X), f_r(X), f_e(X)\}^T \\ s.t.\ g_q(X)\backslash I_q\ g_f(X)\backslash I_f\ g_c(X)\backslash I_c \\ \qquad g_t(X)\backslash I_t\ g_s(X)\backslash I_s \end{cases} \quad (1)$$

Where, $X = (X_1, X_2, X_3 \dots X_n)$ represents the reuse time, resource utilization rate, environmental status and various factors affecting product (Q, F, C, T and S). $g_q(X)$, $g_f(X)$, $g_c(X)$, $g_t(X)$ and$g_s(X)$ are functions of product quality, function, cost, delivery time and service respectively. Iq, If, Ic, It and Is are the quality, function, cost, delivery time and service index constants of products respectively. After obtaining the environmental protection index of the product design, the product can be redesigned according to the design rules, and then the environmental protection index of the new scheme can be calculated. The above process can be concluded that the product design is relatively optimal when the calculated results of the environmental protection index have little difference. Therefore, according to the above evaluation indicators and the product life cycle feature model, the green design of the product can be evaluated to obtain the best solution.

5. Summary

With the continuous enhancement of people's awareness of environmental protection, green products are bound to receive more and more attention from the whole society. Based on the concurrent engineering and the concept of green design, through the integration of information technology, each stage of the product life cycle can be synergistic. At the same time By establishing green assessment indicators and adopting appropriate evaluation methods, the development cycle can be improved, quality can be improved, pollution can be reduced, and the goals of green design and sustainable development can be achieved.

6. References

[1] Edlin N 1991 Management of engineering/design phase *Journal of Construction Engineering and Management* **12** 63-75
[2] Lai KF and Yin GF 2003 *Concurrent Engineering*(Beijing: China Machine Press)
[3] Li XN and Liu X 2013 Research on Product Virtual Prototyping Technology and Design Method for Concurrent Engineering *Advanced Materials Research* **819** 339-43
[4] Hu AW and Fu ZH 2002 The Concept of Green Design for Products *Machine Design and Manufacturing Engineering* **3** 7-9
[5] Fu ZH and Peng YC 2000 Green Design for Products *Machine Design and Research* **16** 10-2
[6] Li XN and Liu HL 2016 Research on PLM-Oriented Product Design Evaluation Based on User Experience *Key Engineering Materials* **693** 1905-9

[7] Liu ZF 2008 *Green design method, technology and application*(Beijing: National Defence Industry Press)

[8] Zeng XH, Liu CH and Hao XH 2007 Research on Evaluation System and Method of Packing Machinery Green Design *Packaging Engineering* **28** 71-3

[9] Huang DZ 2004 Green Design of Engineering Machinery *Water Conservancy& Electric Power Machinery* **26** 32-4

[10] Chen M and Li YJ 2005 Evaluation and Research on Green Chroma of Mechanic Product Design Project *Journal of Liaoning Institute of Technology* **1** 78-82

[11] Chen Z,Liu F and Chen XH 1999 Product Multi-life-Cyele Engineering Based on Green Manufacturing *China Mechanical Engineering* **10** 233-5

Acknowledgments

This work was supported by a grant from the National Social Science Foundation of China (No. 14CG130).

IOP Conf. Series: Materials Science and Engineering **627** (2019) 012007 doi:10.1088/1757-899X/627/1/012007

A Robust Optimization Scheduling for Carrier Aircraft Support Operation Based on Critical Chain Method

Peilong Yuan[1], Wei Xiao[2], Zhengyang Li[1 a], Hongyu Liang[1], Zhenyu Zhao[3]

[1]*Naval Aviation University, Yantai 264001, China;*
[2]*China State Shipbuilding Corporation, Beijing 10094, China;*
[3]*92950 Army Force, Huludao 125100, China*

[a]Corresponding author's e-mail address: lizhengyang1021@163.com

Abstract. The maintenance and service support of carrier aircraft (MSSCA) is a complex process involving many types of resources and activities that require optimisation which can be considered as a multi-resource constrained multi-project scheduling problem (MRCMPSP). It is of great significance to optimize the makespan and obtain a proactive robust schedule to be adaptive to the changes in the dynamic flight deck environment. This paper develops a critical chain method (CCM) for carrier aircraft support robust scheduling in a time-critical and resource-constrained operation environment. The CCM consists of developing a desirable deterministic schedule under multi resource constraints and time-critical issues, and adding a project buffer (PB) to the end of the schedule is used to get a robust proactive scheduling and deal with uncertainty. The triangular fuzzy number (TFN) is applied to describe the durations of each activities and calculate the PB size and obtain an appropriate proportionality between the activity duration and the buffer size. In the numerical research, different size simulations are designed to calculated the robust optimization scheduling. The computational results show that the CMM for carrier aircraft support robust scheduling can make a better decision in the robust proactive optimization scheduling about the resource allocation.

1. Introduction

Scheduling is essential for a wide range of engineering applications such as project management, job shop scheduling and system operations, it is about a process to creating an optimal schedule, which consists of a list of times at which possible tasks, events, or activities are intended to be operated. A poor schedule of a large-scale project may result in huge economic loss to its host company. The presented work was motivated by the need of support operation scheduling general assembly of aircrafts on the aircraft carrier flight deck. Nowadays, aircraft carrier is the largest vehicle sailing on the sea, it is a symbol of the general strength of a military power. The combat capacity of the aircraft carrier is closely correlated with the sortie and recovery abilities of the carrier aircraft on the flight deck, while the sortie and recovery performances of the carrier aircraft depend largely on the scheduling strategies carried out during the deck support operations. The maintenance and service support of carrier aircraft (MSSCA) is the main operation process before the carrier aircraft takes off. On the flight deck, each carrier aircraft requires a series of certain supporting operations in a certain order, such as inspection, oiling, gas supply, weapon mount, and inertial alignment, etc, which are required to be performed within one deck cycle to keep high-intensities sortie rates. Therefore, a proactive baseline schedule for the MSSCA is very necessary before the flight deck operation. On the other hand, the MSSCA problem is a kind of typical dynamic scheduling with many strong constraints, such as time urgency, space restraint and resources restraint. Meanwhile, there exist many uncertainties and disturbances during the scheduling process. It is imperative to make a robust optimization schedule and resource allocation plan for flight deck operations with uncertainty which satisfy the constraints of space, resource, timing and personnel as the efficiency and robustness of flight deck operations are always the bottleneck of improving the sortie generation capacity and operational management level of aircraft carrier.

The keys about the MSSCA are to achieve the aircraft scheduling through different facilities or positions with set limitations in time, space and resource and make the robust proactive schedule be adaptive to the uncertain environment. The research about aircraft carrier deck scheduling for the operations and maintains with limited resources can be considered as resource-constrained project scheduling problems (RCPSP). RCPSP has gained widespread attention for the last few years due to its practical importance and computational challenge. The operation scheduling of MSSCA in the uncertain flight environments, may face many uncertain factors during the flight deck operations, such as stochastic service time of each operation, the resource unavailability, machine breakdowns, operator variation, and even the uncertain change of mission. Moreover, precedence constraints and multiple resource constraints are more complicated in the flight deck operations of carrier, and various kinds of aircraft make the process of flight deck operations diverse. The dynamic scheduling research for MSSCA process can be described as multi-resource constrained multi-project scheduling problem (MRCMPSP) under uncertain environment. Uncertainty in the aircraft carrier usually leads to unscheduled delays, cost overruns, cause the delay and disputes between the operating parties. Managers in the flight deck strongly need an effective method that considers resource constraints and uncertainty for the MSSCA problem, and provides a systematic mechanism for managing schedule during the MSSCA project execution. Therefore, it is imperative to make a time-saving and robust schedule and resource allocation plan for flight deck operations with uncertainty which satisfy the constraints of space, resource, timing and personnel.

Critical chain method (CCM) is a better choice for planning and managing MSSCA that emphasises the resources required to execute project tasks, it is introduced by Goldratt [1] based on the probability theory. CCM has been widely used for project scheduling [2-5]. It uses a deterministic schedule integrated by a buffer mechanism to deal with both resource constraints and uncertainty, and help managers to guarantee the in time and on budget completion of the project. This paper illustrates a robust makespan-based, reliability-based and resource-constraints optimization model for the proactive scheduling of the MSSCA problem based on CCM. In this paper, a project buffer for the MSSCA is placed at the end of the critical chain to protect against exceeding the project deadline, and feeding buffers are placed at the intersections between any non-critical chains and the critical chain to protect it against disturbances. The structure of the paper is as follows. Section 2 presents a literature review of related work about MSSCA and CCM. In the section 3, the MSSCA problems which are consider as MRCMPSP are described in detail. The mathematical programming model for the MSSCA is established. In section 4, the CCM are proposed to deal with the uncertain environment and a buffer is placed to get a robust proactive. In the section 5 demonstrates the model with an example by a proactive robust scheduling optimization algorithm. Finally, we provide overall conclusions in section 6.

2. Literature review

An actual aircraft carrier flight deck environment is one of the most dangerous work environments in the word. a series of researches were conducted, and some research papers have been published about the carrier aircraft scheduling. In the early years, Ouija Board [7] -an aircraft carrier deck simulation platform, was employed by US Navy to monitor the arrangement and sought an optimized schedule by artificial expertise. Ryan [6-7] developed a deck operations course of action planner (DCAP) which was based on human-computer interaction concept and utilizes a conventional integer linear program-based planning algorithm. Then some other research had been published such as the markov decision process model solved by inverse reinforcement learning technique [8], agent based model [9] which was compared with that of the expert user heuristics over a set of experimental scenarios, and queuing network-based policy [10] generator to generate action plan for deck operation. Yu et al [11] established an extended flexible job shop scheduling model for the MSSCA considering the serial and parallel precedence relations between operations and designed an improved differential evolution algorithm (IDE) to improve the computational efficiency. Feng Qiang [12] proposed a multi-agent based fleet maintenance personnel configuration method to solve the mission-oriented aircraft fleet maintenance personnel configuration problem. Wu et al. [13] designed a GERT-based global sensitive method to analyse the uncertainty of flight deck operations. Ryan et al. designed the deck operations course of action planner include the interactive local and global decision support system, expert user heuristics and agent-based model [14-16].

Zhi Zhang [17] designed a human-computer cooperation decision planning system (aircraft carrier deck operation planner (ACDOP) system). Some other research papers are published using the Markov decision process (MDP) to solve the multi-stage decision course under uncertainty on flight deck, the Markov decision process (MDP) based on the reverse reinforcement learning [18-20]. For the uncertain environment in the flight deck, Chao [21] designed a dynamic hierarchical task network (HTN) planning process for the dynamic aircraft carrier flight deck tasting planning. But the two fold network structure is simple and still have some limitations. Jürgen Kuster [22] presented a real time scheduling method in the disruption management of aircraft turnaround, but it is only a partial research. Feng et al. [23] built an improved direct graph to model and designed an improved ant colony optimization algorithm to optimize the carrier aircraft scheduling.

In the MSSCA scheduling which is consider MRCMPSP, some uncertainties and dynamic disruptions exist in carrier aircraft scheduling, such as resource disruptions (resource station breakdown, the operators reduce) and time disruptions (the operational time for the equipment maintenance, which may become longer). The keys to achieving the carrier aircrafts' scheduling include different facilities, operational crews, and positions with set limitations in time, space, and resources. In order to address the uncertainty in MSSCA, CCM which is based on the probability theory is attracted considerable attention. CCM is a proactive robust scheduling method which uses a deterministic schedule integrated by a buffer mechanism to deal with both resource constraints and uncertainty [24]. The CCM can help the flight deck commander deal with both resource constraints and uncertain dynamic disruption in time by a buffer mechanism. It overcomes the deficiency of traditional project methods based on the assumption that there are unlimited resources for the execution of the activities. In the reality project resources are not unlimited. Thus, scheduling without considering resource constraints gives unreliable schedules. Due to the uniqueness of some activities in the MSSCA scheduling, and uncertain dynamic perturbations about activity durations, the project manager in the flight deck may not correctly characterize these random variables. The concept of buffer was adopted which is a more reliable alternative in this situation and can absorb the uncertainty in the MSSCA. The CCM with project buffer provides useful information for the MSSCA scheduling and against unpredictable failures, accidents and delays in the flight deck. Some research and literature had been published to research the types of buffer [25].The CCM has been researched in some problems but not involved in the MSSCA scheduling. On the other hand, the MSSCA scheduling considered as MRCMPSP can not been adequately solved in the dynamic environment. The feeding buffers in the traditional CCM may fail to act as a real robust proactive protection mechanism in the MSSCA scheduling, so the robust optimization scheduling research for MSSCA based on CCM is very necessary and meaningful.

3. Problem description

3.1. Description of MSSCA scheduling problem

The carrier aircraft fleet executes missions in a cyclic way in the flight deck, each aircraft waiting to take off in the flight deck should go through the processes of sortie preparation, launch, flight mission and landing. Once the aircraft is chained in the flight, it will turn to the pre-flight preparation stage called MSSCA. The conduct of MSSCA operations involves the intricate scheduling of aircraft, support equipment and personnel and allocate the supporting resource (renewable resources and non-renewable resources) to complete the operations for preparing the aircraft well before taking off. The resource constrained include the support crew constraint, support equipment constraint, support space constraint, and non-renewable resources constraint. The support crew resource contains different majors (machinery major, ordnance major, avionics major, and special equipment major). The support equipment resource contains different resource stations (oiling station, oxygen and nitrogen station, hydraulic station, power station). The support space resource such as aircraft cockpit make the support activity existing space constraints so that the cockpit inspection of each major cannot be proceed simultaneous. The non-renewable resources include oil, gas, weapon, *etc*. The set of the four types of resources are denoted as *Kp, Ke, Ks* and *Kw* respectively.

In the flight deck, a single aircraft support operation can be considered as a project, and a team of aircraft (denoted as a set $I=\{1,2,...,N\}$) scheduling problem can be considered as multi project scheduling problem. The MSSCA scheduling problem is considered as MRCMPSP in this paper. For each aircraft $i \in I$, a set of real operation activities (denoted as J_i) should be scheduling with serial and parallel constraint relationships. In the flight deck, the supporting operations of each carrier aircraft are about 14 activities, inspections of each part, fuelling, arming, oxygen filling, nitrogen charging, alignment of INS (inertial navigation system), and so on. Besides a virtual start process and a virtual finish process are set to combine the operational activities of each carrier aircraft to a muti-aircraft network flow. The precedence relationships of MSSCA are usually modelled as activity on node network (AON). Figure 1 shows the support operation flow chart of single-aircraft.

Captions should be typed in 9-point Times. They should be centred above the tables and flush left beneath the figures.

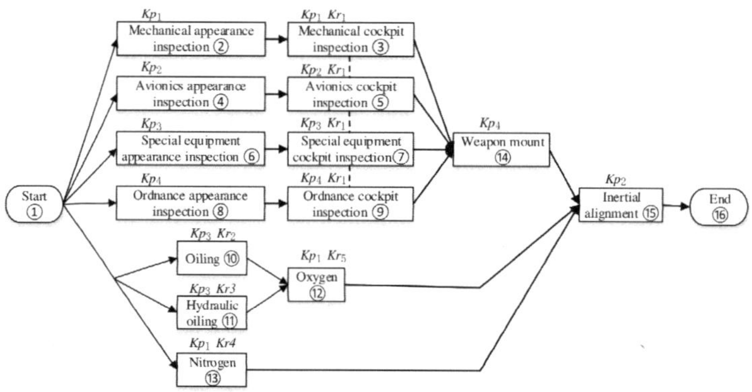

Figure 1. The support process flow chart of each single carrier aircraft

In the Figure 1, the relationships of the MSSCA scheduling process for a single carrier aircraft are as usually modelled as activity on node network *G* (*N*, *P*). O_{ij} is set to denote the *j*th operation of *i*th aircraft, each activity O_{ij} cannot be started until all the operations in the immediate predecessor operation set P_{ij} are completed. Here P_{ij} represents the serial and parallel constraint relationships activities of the *i*th aircraft *j*th activity. As for each operation activities O_{ij}, a set of different kinds of support resources are required to complete the support activities, *Kp*, *Ke*, *Ks* and *Kw* represent different kinds of support resource in the Figure 1. Each operation activities O_{ij} has a different support duration d_{ij}. The support time has a stochastic value because the dynamic scheduling environment, its expectation is relevant to the mission requirement and uncertain environment. Besides, a stationary resource station can only support some settled carrier aircraft. Lr_k is the set of *k*th ($k \in Kr$) supporting resource station. The person in a single major supporting operational crew can support all the aircraft. Lp_k is set of *k*th ($k \in Kp$) supporting operational crew.

3.2. Description of MSSCA scheduling problem based on CCM under uncertainty

In addition to the MSSCA scheduling problem, the scheduling activities are subject to considered uncertainty (e.g., resource station breakdown, the operators reduce and time disruptions because of bad weather and so on). The uncertainty in the scheduling process can lead to the schedule delays, and cost overruns, and dispute between the different scheduling parties. It is very important to calculate and analyse MSSCA activities duration under uncertainty and resource constraints, and provide the more reliable robust schedule to reduce the overrun. In the flight deck, the commander needs to measure the difference of the actual MSSCA schedule and the initial planned schedule so that managerial actions can be made. A robust optimization scheduling method for carrier aircraft support operation is very necessary to estimate the (uncertain) durations of delays/interruptions for the scheduling activities, and their effects on the remaining schedule under resource constraints, so that the proposed method can provide a systematic mechanism for managing schedule during MSSCA project execution. In order to deal with the uncertainty, the fuzzy CMM is used to deal with the continuous relationship between time-resource. It is easy and simple to be applied and incorporated with MSSCA management system and software to take advantage of available resource level functions. In this paper, the triangular fuzzy number (TFN) is applied to describe the durations of each activities.

4. Fuzzy critical chain method for MSSCA robust scheduling

4.1. Solution methodology

The CCM method for MSSCA scheduling can be implemented as follow three interactive and iterative stages:

Step 1: In the first stage, the proactive scheduling strategy is used to create a static schedule using the heuristic algorithm with time/resource take off to provide a minimum-makespan deterministic schedule which will generally be a good starting point for project execution. The starting time and ending time of each activities can be predetermined in the proactive schedule

Step 2: In order to deal with the uncertainty, the proposed method uses fuzzy numbers is designed to model uncertainty in activity durations of MSSCA scheduling based on the deterministic activity durations in the first stage and expert knowledge, a project buffer (PB) is added to deal with uncertainty by a fuzzy set-based schedule. The size of PB is determined by computations with fuzzy numbers.

Step 3: In the third stage for the MSSCA scheduling execution, the proposed CCM is executed to focuses on the penetration level in PB, and then dynamically updates the schedule to provide a more accurate schedule for actual progress in the dynamic scheduling environment.

The three above stages are repeated until the MSSCA scheduling project is completed. The main problem in the robust optimization scheduling for MSSCA is to find a deterministic schedule, which has suitable durations and optimal start times of activities, so that project duration under multi resource constraints is minimised. The proposed method uses fuzzy theory to balance the relationship between them.

4.2. Implementation of MSSCA based on fuzzy theory

The uncertainty of project activity duration generally uses an *L-R* type fuzzy number, The definition of an *L-R* fuzzy number is as follows: *R* is the real number field, If $\overline{A}$ is a normal fuzzy number set of *R*, and the cut set $\overline{A_\varphi}$ of $\overline{A}$ is a closed interval for any $0 \le \varphi \le 1$, then $\overline{A}$ is a fuzzy number. In this paper we select to use the triangular fuzzy number *TFN* (*a*, *b*, *c*) to estimate the uncertainty of MSSCA, as shown in Figure 2. The membership function μ has values in [0, 1], and it is used to express the degree of activity support duration d_{ij} belonging to a fuzzy set. μ is often determined by the carrier aircraft scheduling project manager. The membership function will be as follows:

$$\mu(d_{ij}) = \begin{cases} 0, & d_{ij} < a \\ \dfrac{d_{ij} - a}{b - a}, & a \le d_{ij} \le b \\ \dfrac{c - d_{ij}}{c - b}, & b \le d_{ij} \le c \\ 0, & d_{ij} \ge c \end{cases} \tag{1}$$

Here, a is the minimum duration of activity j for the ith carrier aircraft d_{ij} (a), b is the average duration of activity j for the ith carrier aircraft, c is the maximum duration of activity j for the ith carrier aircraft d_{ij} (c). The support duration d_{ij} can't be in range less than a or greater than c. The support time of each activity is dynamic in the flight deck operation environment. The duration of each activity in different support scenarios is shown in formula (2).

$$d_{ij}(\delta) = d_{ij}(a) + \delta\left[\frac{d_{ij}(c) - d_{ij}(a)}{2^m - 1}\right] \tag{2}$$

Here. $0 \le \delta \le 2^m - 1$, d_{ij} (δ) is the support duration of activity j for the ith carrier aircraft in the different environment. Δ represents different dynamic support environment. The same support activity will cost different duration in different support environment. $\mu(d_{ij})$ is membership function of the support duration d_{ij}. Each activity duration is calculated to get the duration matrix.

$$\begin{bmatrix} d_{11}(1) & d_{12}(1) & \cdots & d_{ij}(1) \\ d_{11}(2) & d_{12}(2) & \cdots & d_{ij}(2) \\ \vdots & \vdots & \vdots & \vdots \\ d_{11}(\delta) & d_{12}(\delta) & \cdots & d_{ij}(\delta) \end{bmatrix} \tag{3}$$

The probability matrix of a random dynamic support operation with the membership μ is as follows:

$$P(d_{ij}(\omega)) = \frac{\mu(d_{ij}(\omega))}{\sum_j^n \mu(d_{ij}(j))}, 1 \le \omega \le \delta \tag{4}$$

The probability matrix can be obtained according to the duration matrix (3) and the probability formula (4).

$$\begin{bmatrix} P_{11}(1) & P_{12}(1) & \cdots & P_{ij}(1) \\ P_{11}(2) & P_{12}(2) & \cdots & P_{ij}(2) \\ \vdots & \vdots & \vdots & \vdots \\ P_{11}(\delta) & P_{12}(\delta) & \cdots & P_{ij}(\delta) \end{bmatrix} \tag{5}$$

A duration vector can be form through selecting each activity duration from the duration matrix.

$$\boldsymbol{d} = (d_{11}(x_1), d_{12}(x_{13}), \cdots, d_{ij}(x_{ij})) \tag{6}$$

Here $1 \le x_1, x_{13}, \cdots x_{ij} \le \delta$, the probability of duration matrix in the corresponding situation can be calculated according to the probability matrix. By repeating the above process, the scenario set and the probability of different dynamic situation can be obtained.

In this paper, the range of activity duration are settled through referring the experience of MSSCA project participants and the judgment. The personnel behaviour factors still play a role in the scheduling project. Therefore, the application of fuzzy theory in the critical chain technology still follows the above hypothesis.

4.3. Identifying critical chains and project buffer

A chain in the MSSCA is a sequence of support activities under both precedence and resource dependencies. In the robust optimization scheduling for carrier aircraft support operation based on CCM, the critical chain is the longest chain, and it will determine the project duration. In the scheduling project, if there are many critical chains can be selected, a critical chain with greatest uncertainty will be scheduled.

The proposed method deals with the dynamic scheduling environment by adding a project buffer (PB) at the end of the selected critical chain. We selected to add the project buffer to the end of the chain because that the resource conflicts often happen during the transfer of the actual information from activities on the chain to the end of the chain in the dynamic scheduling environment. In dynamic scheduling environment, the support finished completion time is not only depended on the sort order of the scheduling activities, but also depend on the duration of each scheduling activity. The size of PB is determined by computations with fuzzy numbers. The σ_m is defined as the variance of an activity duration in the critical chain under different scenarios. The PB is calculated as follows:

$$PB = \sqrt{\sum_{i,j \in C} \sigma_{ij}^2} \tag{7}$$

5. Robust optimization scheduling model for MSSCA based on CCM

In the MSSCA scheduling model, some other sets, parameters and decision variables are set as follows:

Rp_{kl} is the set of the carrier aircraft support range by the k_{th} ($k \in Kp$) major lth ($l \in Lp_k$) operational crew.

Re_{kl} is the set of the carrier aircraft support range by the k_{th} ($k \in Kr$) type lth ($l \in Lr_k$) station.

SX_i start time for supporting the ith carrier aircraft

S_{ij} start time of activity j for the ith carrier aircraft

rp_{ijk} demand number of kth operational crew for the ith carrier aircraft jth activity

rs_{ijk} demand number of kth space for the ith carrier aircraft jth activity

re_{ijk} demand number of kth equipment for the ith carrier aircraft jth activity

rw_{ijk} demand number of non-renewable resources for the ith carrier aircraft jth activity

$Cmax$ the maximum completion time of the activities

Xp_{ijkl} 1 if the kth station lth equipment is allocated to activity j for the ith carrier aircraft; 0 otherwise

Xe_{ijkl} 1 if the kth station lth equipment is allocated to activity j for the ith carrier aircraft; 0 otherwise

x_{ijk} 1 if the activity j for the ith carrier aircraft has been finished at the t moment; 0 otherwise

The mathematical description of the problem based on CMM is given as follows:

Objective function.

$$\text{Min } F = \min\,(C_{max}+PB) \tag{8}$$

Constraints.

$$S_{i1} \geq SX_i, \forall i \in I \tag{9}$$

$$E_{ij} = S_{ij} + d_{ij}, \forall i \in I, \forall j \in J_i \tag{10}$$

$$S_{ij} \geq S_{ih} + d_{ih}, \forall (i,h) \in P_{ij}, \forall i \in I, \forall j \in J_i \tag{11}$$

$$\sum_{O_{ij} \in I_{it}} rs_{ijk} \leq 1, \forall i \in I, \forall j \in J_i, \forall k \in Ks \tag{12}$$

$$\sum_{i=1}^{n} \sum_{j \in J(t)} rp_{ijk} \leq |Lp_k|, \forall k \in Kp, t \in [0,T] \tag{13}$$

$$\sum_{i=1}^{n}\sum_{j\in J(t)} re_{ijk} \le |Lr_k|, \forall k \in Kr, t \in [0,T] \quad (14)$$

$$\sum_{i=1}^{n}\sum_{j\in J(t)} rw_{ijk} \le |Lw_k^t|, \forall k \in Kw, t \in [0,T] \quad (15)$$

$$\sum_{l\in Lr_k} Xe_{ijkl} = re_{ijk}, \forall i \in I, \forall j \in J_i, \forall k \in Kr \quad (16)$$

$$\sum_{l\in Lp_k} Xp_{ijkl} = rp_{ijk}, \forall i \in I, \forall j \in J_i, \forall k \in Kp \quad (17)$$

$$\sum_{i\in I-Re_{kl}}\sum_{j\in J_i} Xe_{ijkl} + \sum_{i\in I-Rp_{k'l'}}\sum_{j\in J_i} Xp_{ijkl} = 0, \forall j \in J_i, \forall k \in Kr, \forall l \in Lr_k, \forall k' \in Kp, \forall l' \in Lp_k \quad (18)$$

The whole project schedule consists of two parts, the duration and the buffer. For any given schedule, if the buffer size is too big, the activity duration will be very short. In this case, it is difficult to finish the activity on time. However, if the buffer size is too small, it may not be succeed in dynamic scheduling environment. The above objective function with duration and project buffer is designed to balance the optimal of makespan and the robustness, and avoided the occurrence of the situation that the total makespan is very short and the duration volatility is very uncertain. Constraint 9 to 19 describe the MSSCA scheduling model in the flight deck. Constraint (9) ensures that the first support activity must start before the carrier is transported up to the flight deck and parked in a permanent position. Constraint (10) defines the relationship between the finish time of each activity with the start time. Constraints (11) ensures the precedence relationship, which means that any activity cannot be started before the completion of its preceding activity. Constraint (12) ensures that each activity is executed only once from one exact period, *t*. Constraints (13), (14), and (15) represent the capacity constraints for renewable resources (operational space, operational crews, resource stations), respectively. Constraint (9) represents the capacity constraints for the non-renewable resources in each time period. Constraints (10) and (11) represent the allocation of the operational crews and resource stations, which need to be equal to the demand. Constraint (12) represents the support coverage of the operational crews and resource stations.

6. Numerical example

In order to test the performance of the new model and the proactive robust optimization method for the MSSCA, a set of simulation scenarios are generated to make the experimental research. Double population genetic algorithm (DPGA) is designed to improve the quality and diversity of the population. First, encoding and decoding processes are set to get a population and matching scheduling. The left and right populations are processed simultaneous with crossover and mutation and local search processes. For the left population, right scheduling is made in the encoding and decoding processes. For the right population, left scheduling is made in the encoding and decoding processes. The original coding is corrected by decoding the end time of the scheduling solution and the elite population is reserved. The pseudo code of DPGA with local search strategy can be found in our previous paper.

Through the adoption of proper buffer sizing methods, the uncertainty factors of the project will be effectively taken into consideration and absorbed. The project buffer also protects against uncertain failures, accidents and delays. The proposed method uses MATLAB software to simulate the uncertainty of the project, the project buffer, the actual duration and cost of the project. The experiments were performed using a PC with two Intel(R) Core (TM) i5-4258U CPU @ 2.4 GHz processors and 4 GB RAM (Intel cooperation, Santa Clara, CA, USA. The algorithm has been coded in MATLAB 2017(a) software (The MathWorks, Inc., Natick, MA, USA).

In the numerical research, the numbers of each professional operational crew are respectively $|K_{p1}| = 8$, $|K_{p2}| = 6$, $|K_{p3}| = 8$, and $|K_{p4}| = 6$. Five kinds of resource stations including the power station (K_{r1}), oiling station (K_{r2}), hydraulic station (K_{r3}), nitrogen station (K_{r4}), and oxygen station (K_{r5}) for the service support operations of non-renewable resources. The numbers of each resource station in the experiment are respectively$|K_{r1}| = 8$, $|K_{r2}| = 6$, $|K_{r3}| = 5$, $|K_{r4}| = 5$, and $|K_{r5}| = 5$. Kw represents the upper limit of non-renewable resources for the simultaneous support operation. Kw is set [6, 5, 2, 4, 2]. Table 1 shows the station set of each kind of resource station that covers the waiting support carrier aircraft in the fixed parking position.

Four kinds of support operational crews and five types of resource stations make up the flexible and sustained carrier aircraft support activities in the flight deck. The average duration of activity j for the ith carrier aircraft b (in minutes) are shown in the Table 2. The maximum duration (c) of each activity and minimum duration (c) of each activity is depend on the average duration, as is shown in the table 2. $a=b-\Delta ab$, $c=b+\Delta bc$.

Table1. Coverage relationship between resource station and carrier aircraft.

NO.	Resource Station Coverage Category				
	K_{r1}	K_{r2}	K_{r3}	K_{r4}	K_{r5}
1	[1]	[1]	[1]	[1]	[1]
2	[1, 2]	[1]	[1]	[1]	[1]
3	[1, 2]	[1, 2]	[1, 2]	[1, 2]	[1, 2]
4	[2, 3]	[2]	[1, 2]	[1, 2]	[1, 2]
5	[3, 4]	[2]	[2]	[2]	[2]
6	[3, 4]	[3]	[2, 3]	[2, 3]	[2, 3]
7	[4, 5]	[3, 4]	[3]	[3]	[3]
8	[5, 6]	[3, 4, 5]	[3, 4]	[3, 4]	[3, 4]
9	[5, 6, 7]	[4, 5]	[3, 4]	[3, 4]	[3, 4]
10	[6, 7]	[5, 6]	[4, 5]	[4, 5]	[4, 5]
11	[7, 8]	[6]	[4, 5]	[4, 5]	[4, 5]
12	[8]	[6]	[5]	[5]	[5]
13	[8]	[6]	[5]	[5]	[5]

Three instances are proposed in the research. **Instance 1:** Five carrier aircrafts（Nos. 2, 3, 9, 10, 12), n = 5. **Instance 2:** Nine carrier aircrafts (Nos. 2, 3, 4, 5, 8, 9, 10, 11, 12), n = 9. **Instance 3:** 13 carrier aircrafts (Nos. 1–13), n = 13. The objects are the buffer, actual duration of the the average, max, min results, and variance, which are calculated with different wave carrier aircrafts. Table 3 shows the results of the MATLAB simulations of these three types of projects with different size. Figure 2 and figure 3 shows the gantt chart of optimal personnel allocation and station allocation in deterministic mission instance 2. The vertical axis represents the number of the supporting operational crew and resource station. $L_{p_k}^l$ represents the lth operational crew of the kth major, $L_{r_k}^l$ represents the lth support equipment of the kth resource station, and $I_i - j$ represents the ith carrier aircraft number for the jth activity. Table 3 shows the results of the MATLAB simulations of these three types of scheduling projects with different size.

Table 2. Support time of each activity in a single carrier aircraft.

	b(min)													*a(min)*	*c(min)*
Act	1	2	3	4	5	6	7	8	9	10	11	12	13		
②	12	13	13	13	13	13	14	14	14	14	14	12	12	$b_{i2}-2.5$	$b_{i2}+2.5$
③	3	4	4	4	4	4	5	5	5	5	5	3	5	$b_{i3}-1.0$	$b_{i3}+1.0$
④	5	3	3	3	3	3	4	4	4	4	4	5	3	$b_{i4}-0.5$	$b_{i4}+0.5$
⑤	7	6	6	6	6	6	5	5	5	5	5	7	5	$b_{i5}-1.0$	$b_{i5}+1.0$

⑥	6	4	4	4	4	4	3	3	3	3	3	5	4	$b_{i6}-0.5$	$b_{i6}+0.5$
⑦	9	7	7	7	7	7	6	6	6	6	6	10	7	$b_{i7}-1.5$	$b_{i7}+1.5$
⑧	0	9	9	9	9	9	10	10	10	10	10	7	0	$b_{i8}-2.0$	$b_{i8}+2.0$
⑨	0	4	4	4	4	4	5	5	5	5	5	3	0	$b_{i9}-0.5$	$b_{i9}+0.5$
⑩	18	13	13	13	10	10	12	12	10	10	10	12	18	$b_{i10}-2.5$	$b_{i10}+2.5$
⑪	4	5	5	5	5	5	5	5	5	5	5	4	5	$b_{i11}+1.0$	$b_{i11}+1.0$
⑫	5	3	3	3	3	3	3	3	3	3	3	4	5	$b_{i12}-0.5$	$b_{i12}+0.5$
⑬	5	3	3	3	3	3	3	3	3	3	3	5	5	$b_{i13}-0.5$	$b_{i13}^{*}+0.5$
⑭	0	12	12	12	12	12	11	11	11	11	11	9	0	$b_{i14}-2$	$b_{i14}+2$
⑮	10	7	7	7	7	7	6	6	6	6	6	6	6	$b_{i15}+1.5$	$b_{i15}+1.5$

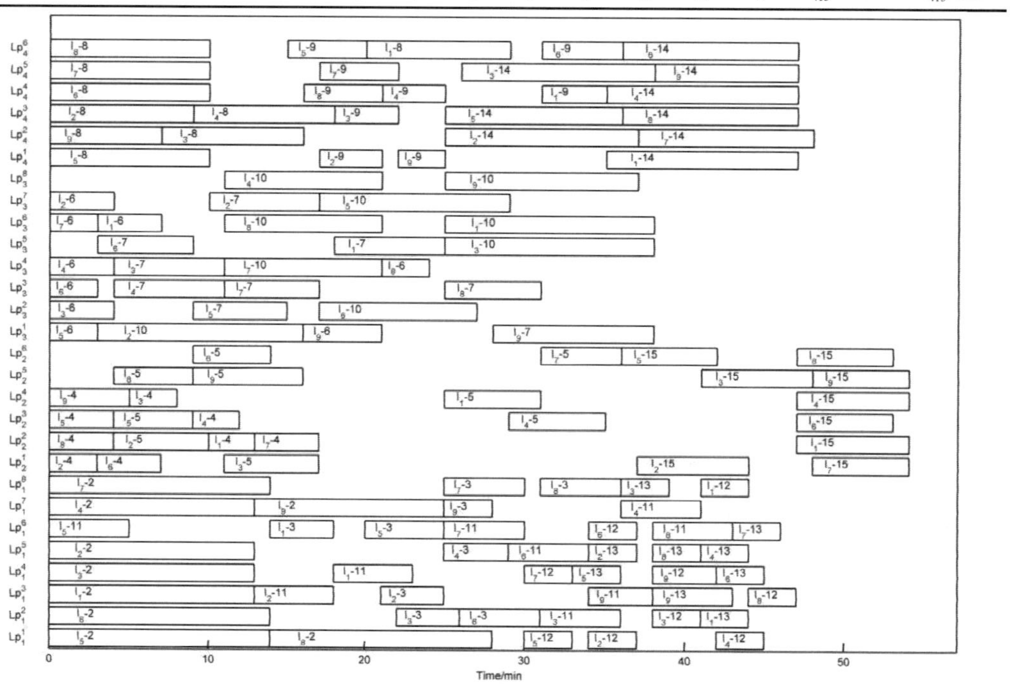

Figure 2. Gantt chart of optimal personal crews in instance 2

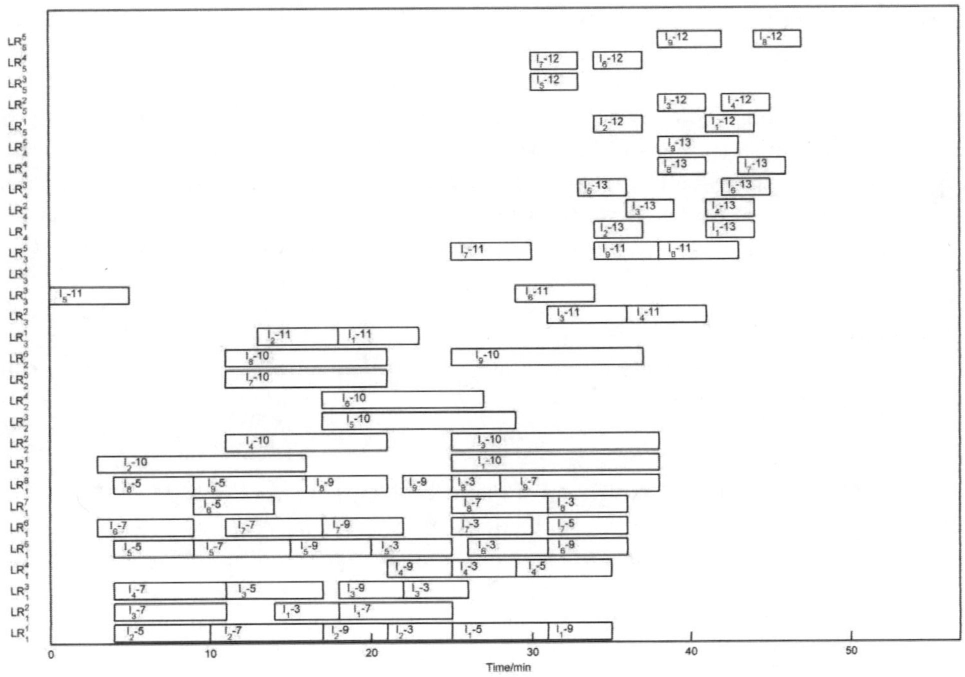

Figure 3. Gantt chart of optimal resource stations in instance 2

Table 3. The results of Matlab simulation.

Project size	PB(min)	Actual duration (min)			*Variance*
		Average	Max	Min	
Instance 1	5	43	45.5	40.5	2.053
Instance 2	9	54	51	56.0	3.246
Instance 3	11	65	68.5	61.5	4.102

In the table 3, the project buffer, mean value, maximum value, minimum value and the standard deviation of the actual duration and simulated cost were compared with different scheduling instance. The project buffer determined by both CMM and the project size is a fixed value, and we take the average value of the 20 simulations as the project buffer of the proposed method. We see that the buffer consumption, the actual duration of the proposed CMM method and TFN increase with the increase in the degree of project size. The variance of the results also increases in the 20 simulations. Based on the results of these cases, we have reached that the buffer size is determined by the proposed method and the project size. It may make the buffer size increase linearly with the number of activities, resulting in an excessive buffer size. The proposed method of CMM with TFN can reduce the uncertainty in project duration and provide convenient guidance to carrier aircraft support and scheduling applications in the dynamic flight deck environment, and thus reduce the buffer. Therefore, it could further reduce the duration fluctuations and improve the on-time completion rate.

7. Conclusion

This paper presents a proactive robust optimization method named CCM based on fuzzy method for the MSSCA that considers as a comprehensive multi resource-constrained multi project scheduling problem. A triangular fuzzy number is used to calculate the resource requirement and considers the comprehensive factors affecting the resource constraints to adapt uncertainty scheduling environment. The MSSCA scheduling project consists of two parts, the duration and the buffer. An integer comprehensive mathematical formulation for MSSCA is established with the robust objectives of the limitative makespan and the project buffer by considering the independent operational time and resource constraint of each activity. Simulation results shows that the proposed method can make a better robust proactive scheduling for the MSSCA project. In the MSSCA proactive scheduling, if the project buffer size is too big, the activity duration will be very short. It will be difficult to finish the operation activities on time. If the buffer size is too small, the proactive scheduling will not be overcome completely in the dynamic environment. In the future, we can make an important extension regarding research about the developed robustness scheduling method for the project buffer to improve with the proposed resource allocation schemes and adapt the dynamic flight deck environment.

Acknowledgments
The authors highly appreciate the valuable comments provided by the reviewers, whose comments contributed significantly to improve the paper.

References
[1] Goldratt E. Critical chain. The North River Press Publishing Corporation; 1997. 246p.
[2] Ahlemann F, El Arbi F, Kaiser M G, et al. A process framework for theoretically grounded prescriptive research in the project management field[J]. International Journal of Project Management, 2013, 31(1):43-56.
[3] Long L D, Ohsato A. Fuzzy critical chain method for project scheduling under resource constraints and uncertainty[J]. International Journal of Project Management, 2008, 26(6): 688-698.
[4] Li J, Wang R, Yang Y. A New CCPM(Critical Chain Project Management)Scheduling Algorithm Based on Resource Conflict Solution[J]. Journal of North western Polytechnical University, 2010, 28(4):547-552.
[5] Leach L P. Critical chain project management improves project performance[J]. Project Management Journal, 1999, 30(2): 39-
[6] Ryan J C, Cummings M L, Roy N, et al. Designing an interactive local and global decision support system for aircraft carrier deck scheduling[C]//Proc. AIAA Infotech@ Aerospace Conf. 2011
[7] Ryan J C, Banerjee A G, Cummings M L, et al. Comparing the performance of expert user heuristics and an integer linear program in aircraft carrier deck operations[J]. IEEE transactions on cybernetics, 2014, 44(6): 761-773.
[8] Michini B, How J P. A human-interactive course of action planner for aircraft carrier deck operations[J]. AIAA, 2011, 201(1): 1.
[9] Ryan J C, Cummings M L. Development of an agent-based model for aircraft carrier flight deck operations[J]. M&S Journal, 2014, 2014: 5-15.
[10] Dastidar R G, Frazzoli E. A queueing network based approach to distributed aircraft carrier deck scheduling[J]. AIAA, 2011, 1514.
[11] L Yu, C Zhu, J Shi et al, "An extended flexible job shop scheduling model for flight deck scheduling with priority, parallel operations, and sequence flexibility," Scientific Programming, vol. 2017, no. 1, pp. 1-15, 2017.
[12] Feng Q, Li S, Sun B. A multi-agent based intelligent configuration method for aircraft fleet maintenance personnel[J]. Chinese Journal of Aeronautics, 2014, 27(2): 280-290.
[13] Y Wu, X Pan, R Kang et al, "Multi-parameters uncertainty analysis of logistic support process based on GERT,"Journal of Systems Engineering & Electronics, 2014, 25(6): 1011-1019, .
[14] Ryan J C, Cummings M L, Roy N, et al. Designing an interactive local and global decision support system for aircraft carrier deck scheduling[C]//Proc. AIAA Infotech@ Aerospace Conf. 2011.
[15] Ryan J C, Banerjee A G, Cummings M L, et al. Comparing the performance of expert user heuristics and an integer linear program in aircraft carrier deck operations[J]. IEEE transactions

on cybernetics, 2014, 44(6): 761-773.

[16] Ryan J C, Cummings M L. Development of an agent-based model for aircraft carrier flight deck operations[J]. M&S Journal, 2014, 2014: 5-15.

[17] Zhang Z, Lin S, Dong R, et al. Designing a Human-Computer Cooperation Decision Planning System for Aircraft Carrier Deck Scheduling[M]//AIAA Infotech@ Aerospace. 2015: 1111.

[18] B Michini and J P How, "A human-interactive course of action planner for aircraft carrier deck operations," in AIAA Infotech@ Aerospace 2011, 1515, 2011.

[19] Michini, B.; How, J. P. A human-interactive course of action planner for aircraft carrier deck operations. AIAA Infotech@ Aerospace, 2011.

[20] Li, Y., Zhu, Y., Yang, F., & Jia, Q. Inverse reinforcement learning based optimal schedule generation approach for carrier aircraft on flight deck. Journal of National University of Defense Technology, 2013, 35(4): 171-175.

[21] Qi C, Wang D. Dynamic Aircraft Carrier Flight Deck Task Planning based on HTN[J]. IFAC-Papers OnLine, 2016, 49(12): 1608-1613.

[22] Jürgen Kuster, Dietmar Jannach. Handling Airport Ground Processes Based on Resource-Constrained Project scheduling[J]. Springer Berlin Heidelberg , 2006 , 4031 :166-176

[23] Q Feng, W. Bi, B. Sun et al, "Dynamic scheduling of carrier aircraft based on improved ant colony algorithm under disruption and strong constraint," Second International Conference on Reliability Systems Engineering, 2017. 1-9.

[24] Lim, A., H. Ma, B. Rodrigues, S. T. Tan, and F. Xiao. "New Concepts for Activity Float in Resource-constrained Project Man-agement." Computers & Operations Research, 2011. 38 (6): 917–930.

[25] Steyn, H."Project Management Applications of the Theory of Constraints beyond Critical Chain Scheduling." International Journal of Project Management, 2002, 20 (1): 75–80

Research on Changed Design of Gate Position of Window Frame of Automobile Front Door

Haorui Zhai[1],Xin Xing[1],ZhuAn Zheng[1],Conggui Sun[2],Chen Teng[1]

1School of Automobile Engineering, Yancheng Institute of Technology, Jiangsu Yancheng China
2Jiangsu Yuyan Mould Technology Co., Ltd. Jiangsu Yancheng China

18251423907@163.com, Yancheng institute of technology, 1 middle road, hope avenue, tinghu district, yancheng city, jiangsu province, China

Abstract. The front door sash of the car is selected as the research object, and the design of the original gate position causes the warpage displacement of the front door sash to be large. Based on the CAE principle, using Moldex3D mold flow software analysis, it was found that the cause of the large amount of warpage deformation was flow imbalance. Then, according to the principle of flow balance, the design of the gate position was changed. The results of the mold flow analysis showed that the warpage displacement decreased from 18.5 mm to 14.1 mm, and the overall optimization was 24%. And the comparison found that the molding pressure was reduced from 224 MPa to 171 MPa (single cavity), and the clamping force was reduced from 2003 Ton to 1474 Ton (single cavity).

1. Introduction

The window frame of automobile front door is a kind of window frame installed on the front door. It is a kind of exterior ornament with beauty and practicability. Besides making automobile beautiful, it also has good ability of shock absorption and noise elimination[1]. The example product in this paper is an injection moulding part of window frame of front door for luxury car made by Jiangsu Yuyan Mould Technology Co., Ltd. As shown in Fig.1 (a), the product is shaped as "U" and made of PA6 + GF15. It is a product of polyamide including 15% glass fiber, adopting a mode of two points to produce. Its single cavity volume is 302 cc, and the meat thickness of the product is evenly distributed with an average thickness of 1.5 mm. Just like the Fig.1 (b):

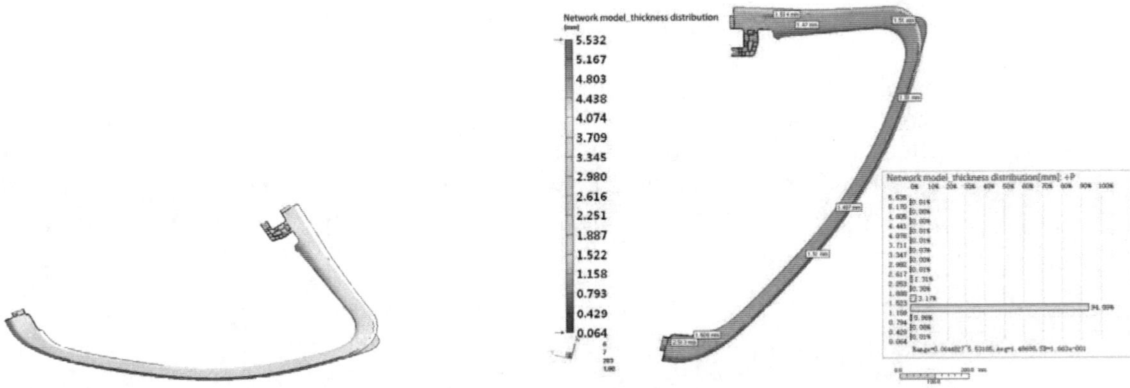

(a) Product model (b) Meat thickness distribution

Fiugure 1. Product parameters

Reducing vehicle weight by using lighter materials is a popular approach for improving fuel efficiency. Due to their ability to enhance mechanical performance as well as satisfy safety and durability requirements, fiber-reinforced-thermoplastic compounds are now commercially available as candidate materials. In general, the shell-core structure, as the orientation structure of the flow-induced orientation of fibers, is found in injection molded parts[2].

The gate of injection mold is the narrow part between the runner and the cavity. It accelerates the movement of the plastic conveyed by the runner, so that the plastic can enter the cavity faster and better.[2] The size and position of gate have a great influence on the quality of plastic parts. The shape of this model is "U". The inappropriate location of gate will lead to unbalanced flow in filling process, which will easily lead to large warpage displacement. The size of gate can be changed according to the factors after opening, but the position of gate is difficult to change after opening, so the position of gate should be analyzed in advance. Based on the finite element model flow analysis technology, product problems can be pre-judged before they arise. Moldex3D Mould Flow Analysis Software developed by Taiwan Kesheng Company is a very useful analysis tool. Its BLM solid grid and hot runner module provide reliable guarantee for the accuracy of Mould Flow Analysis. The simulation results are for four stages of a typical injection moulding: Filling, Packing, Cooling and Warpage[3-5].

2. Analysis of original scheme design

2.1. Design of original gate location

The design of the two points is adopted. The number of grids is too large, and the analysis time is longer. Based on the symmetric cavity, the analysis of single module hole is carried out in order to save the analysis time. The appearance of the product is strictly required, the edge line should be clear and there should be no runner trace, the warpage displacement should not be too large, and the forming period should be controlled within 20 seconds. In view of the above situation, the hot runner technologies adopted, and the gate position is gummed in the way shown in Fig.2 (1)The method of "hot runner + cold runner" is adopted. As shown in Fig.2 (a), the red arrow in the figure represents the hot runner, and the cold runner is used between the hot runner and the plastic part[6]. This design avoids the problems of single hot runner feeding resulting in tiger skin pattern on plastic parts and long product cycle caused by cold runner; (2)The size of gate is 10mm *5mm by side feeding as shown in Fig.2 (b).

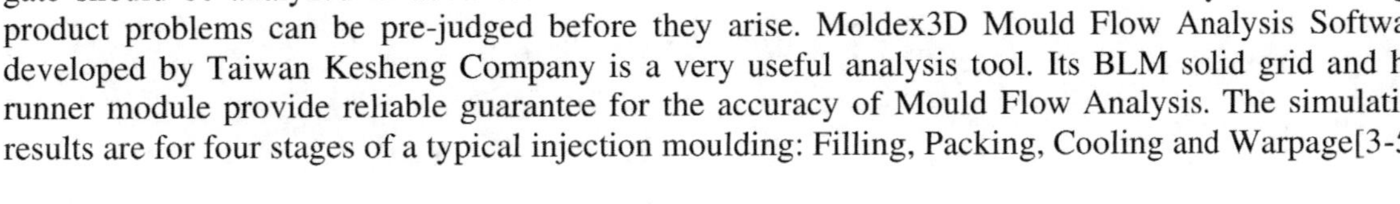

(a) Runner (b) Gate

Figure 2. Glue feeding mode

2.2. Model flow analysis results

The process parameters are set as follows: (1) Filling stage: the filling time is 4sec, and the upper limit of pressure is 200MPa. (2) Pressure holding stage: Pressure holding time is 6 sec, divided into three stages: 4sec/124MPa, 1sec/93MPa, 1sec/46MPa. (3) Cooling stage: cooling time is 10 sec. (4) Temperature setting: Nozzle temperature is 265℃, cooling water temperature is 75℃. The above scheme is simulated and analyzed by Moldex3D. The results are as follows Fig.3.

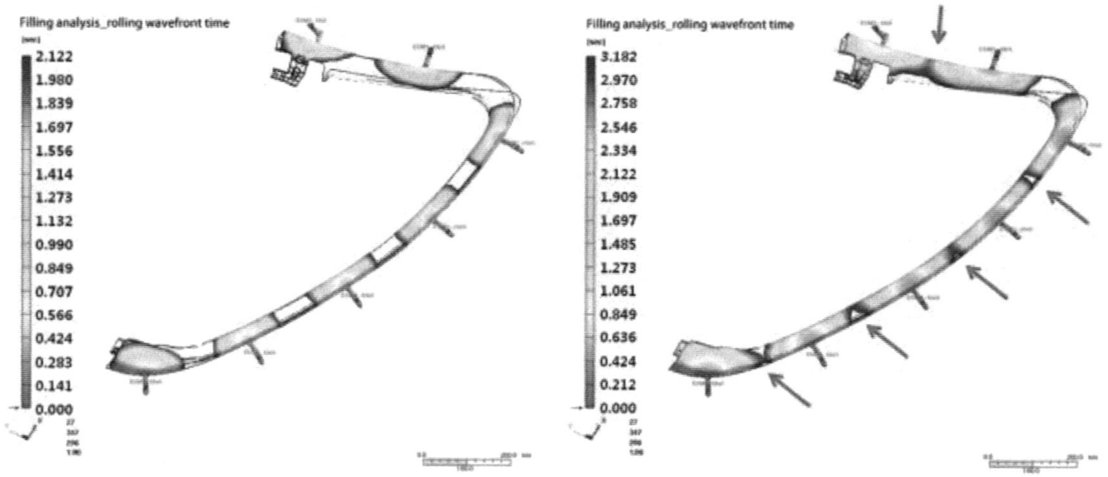

(a) Filling 50% (b) Filling 75%

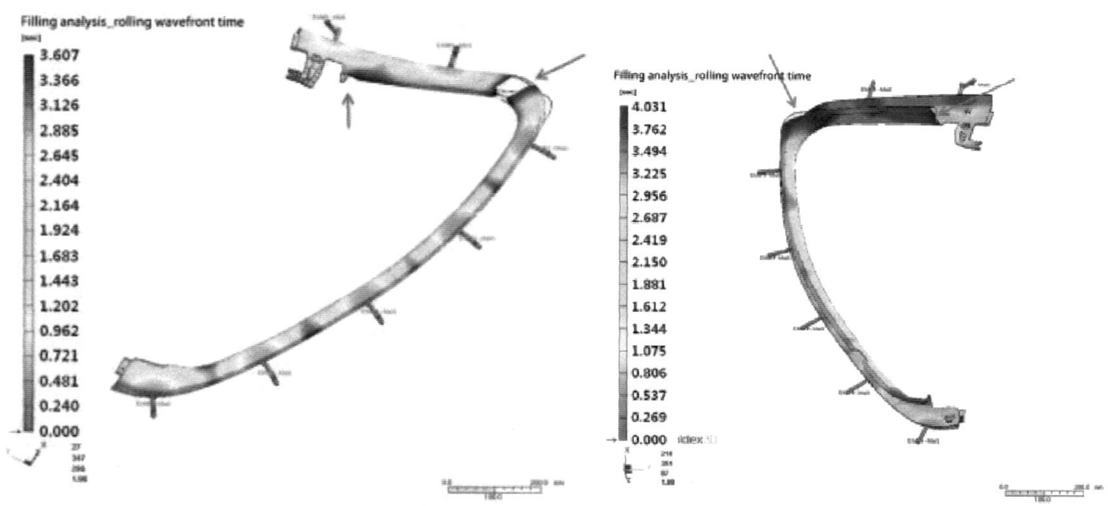

(c) Filling 85% (d) Filling 95%

Fiugure 3. Filling Results

Fig.3 (a) shows that there is no difference when filling 2.12 sec and plastic filling 50%. Fig.3 (b) shows that when filling 3.18sec and plastic filling 75%, the plastic entering the cavity through different gates reaches the flow end at the position indicated by the red arrow on Fig.3 (b). By filling 85% of Fig.3 (c), it can be seen that the flow end at the upper right corner is larger than that at the other flow ends, which indicates that the time of the flow end at the upper right corner of the plastic part is longer than that at other parts. Comparing Fig.3 (b) with Fig.3 (c), it is obvious that the position indicated by the red arrow on the upper right side of the plastic part (c) is longer than the filling time at other parts of the plastic part, which results in unbalanced flow. Fig.3 (d) In filling 4.03 sec, 95% of filling can be seen. The red arrow points to the filling end of the whole plastic part. The filling time here is obviously longer than other parts of the plastic part. It is the last filling and the latest entry into the holding stage, resulting in the phenomenon of flow imbalance.

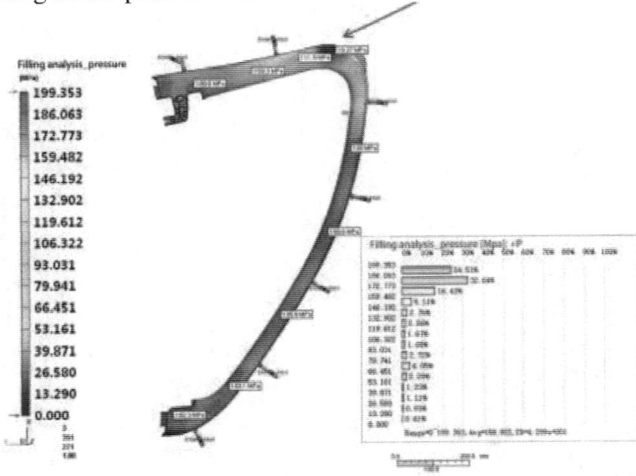

Fiugure 4. Pressure distribution at the end of filling

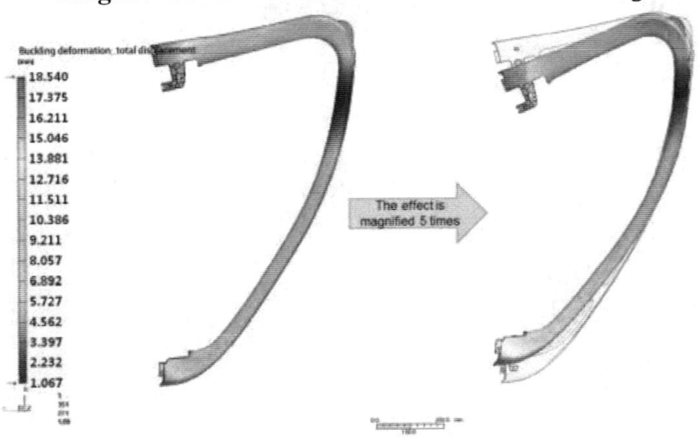

Fiugure 5. Total warpage displacement

The pressure distribution at the end of filling shown in Fig. 4 shows that the pressure rises rapidly in the area where the filling is completed earlier. When the filling is completed, the pressure distribution is not uniform. The red arrow on the graph is the filling end, so the pressure is the lowest. In summary, the color bar in Fig.3 and Fig.4 shows that the time to reach the end of the flow is quite different, and the pressure in the area where the filling was completed earlier rises faster, resulting in uneven distribution of the overall pressure, showing the phenomenon of flow imbalance, which buries a hidden danger for the large warpage displacement.

Fig.5 shows that the maximum warpage displacement of demoulding product is 18.5mm. After enlarging 5 times on the right side, we can see that the shape of the plastic part warps in the direction indicated by the red arrow, and the displacement is large. By comparing the warpage displacement caused by various factors, Fig.6 (a) is the total warpage displacement of 18.5 mm, Fig.6 (b) is the total area shrinkage effect of the displacement of 14 mm, Fig.6 (c) is the fiber alignment effect of the displacement of 9.4 mm, and Fig.6 (d) is the temperature difference effect of the displacement of 0.3 mm. The warpage of plastic parts is mainly due to the uneven shrinkage caused by unbalanced flow in the filling process and the alignment effect of fibers caused by 15% glass fiber in the material, while the warpage caused by temperature difference can be neglected.

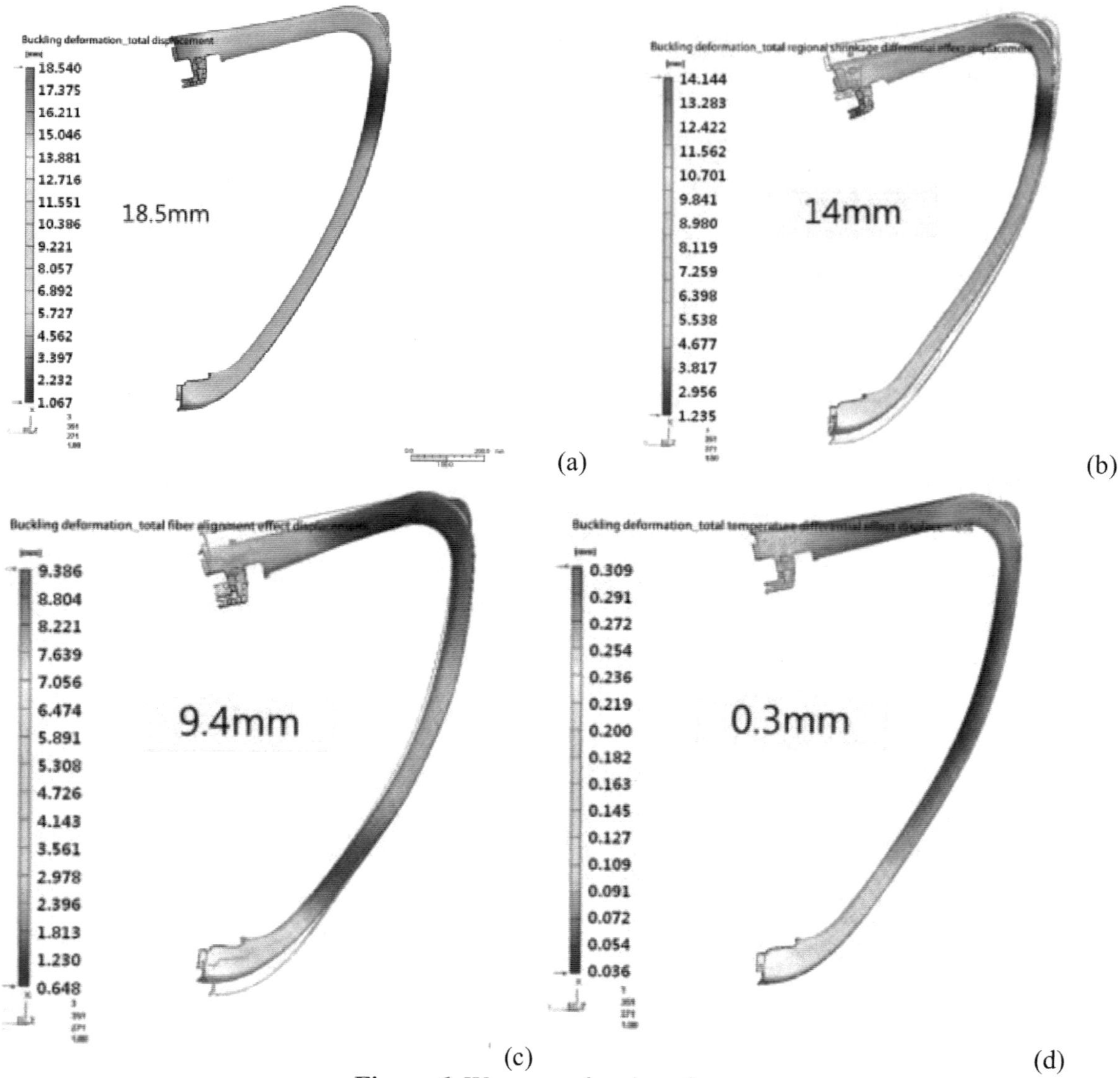

Figure 6. Warpage of various factors

3. Analysis of Optimized Scheme Design

3.1. Gate Position Change Design

Regional shrinkage and fiber matching lead to warpage deformation of 18.5mm of final product, improve flow balance, reduce the pressure imbalance of plastic parts, and effectively reduce the deformation caused by regional shrinkage. The influence of fibers on warpage is more complex. It is usually possible to change the arrangement of fibers only by changing the flow trend, so as to improve warpage. According to the analysis results of the previous mould flow, it is determined that the original gate position design caused the imbalance of flow, resulting in the problem of large warpage displacement of plastic parts, which is difficult to meet the requirements of product quality and the appearance quality of plastic parts, and then the gate position is changed. In order to improve the uneven pressure distribution caused by flow imbalance, the excessive forming pressure and clamping force, and the warping of regional contraction effect caused by subsequent non-uniform shrinkage, the four gate positions at the arrow in Fig.7 are moved towards the flow end, i.e. the direction indicated by the red arrow, to try to adjust the flow.

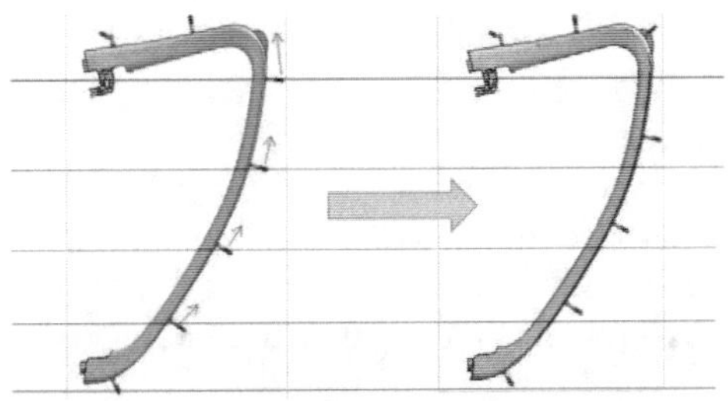

Figure 7. Gate position change design

3.2. Result analysis and comparison

After changing the design of gate position, the result is analyzed by Moldex3D, and the analysis result is shown in Fig.8 (left: original design; right: changed design; left and right are omitted below).

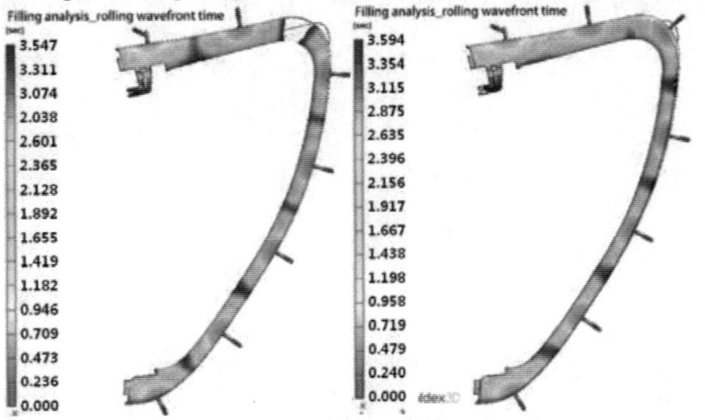

(a) Filling 85%

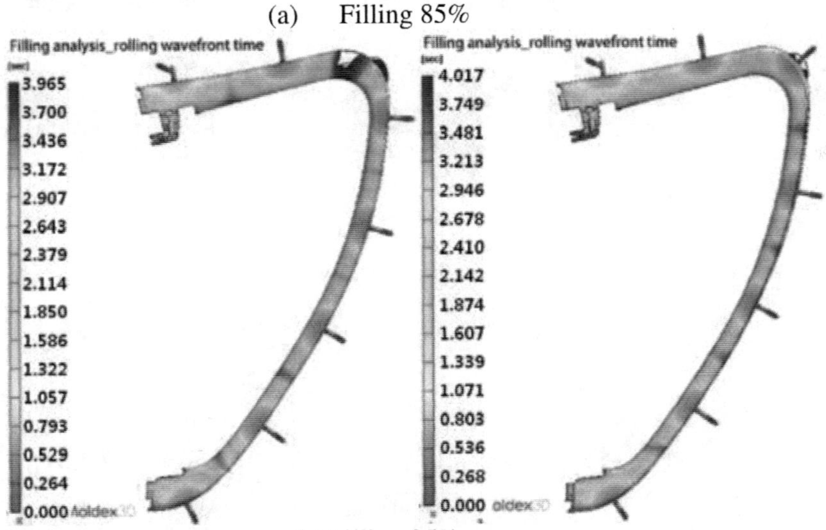

(b) Filling 95%

Figure 8. Result analysis

From Fig.8 (a) 75% filling time, it can be clearly seen that the filling time in the left picture is 75% filling time, the upper left flow end is mostly not full, while the upper right picture is almost full. Fig.8 (b) filling 95% at the end of the flow, through the color bar, we can see that the upper right part of the left picture is still not filled with the cavity, the difference of the flow end time is large, which leads to the phenomenon of flow imbalance; while in the right picture, the cavity has almost been filled, the distribution of the color bar is more uniform, and the flow end time is roughly the same, the difference is not big. By comparing the pressure curves of the two schemes shown in Fig.9 (a) (red curve is the original design curve, green curve is the change design curve, and omitted as red curve and green curve below), it can be seen that the maximum pressure of the red curve is 224 MPa (single die hole), while the maximum pressure of the green curve is 171 MPa (single die). Compared with the original design, the design changes optimize the flow balance. At the same filling speed, the filling pressure is smaller and the forming pressure is reduced by 53 MPa. Comparing the lock force curves of the two schemes shown in Fig.9 (b), it can be seen that the maximum lock force of the red curve is as high as 2003 tons (single die hole), while the maximum lock force of the green curve is only 1474 Ton (single die hole). The smaller the forming pressure, the smaller the lock force requirement, the lower the lock force by 529 Ton.

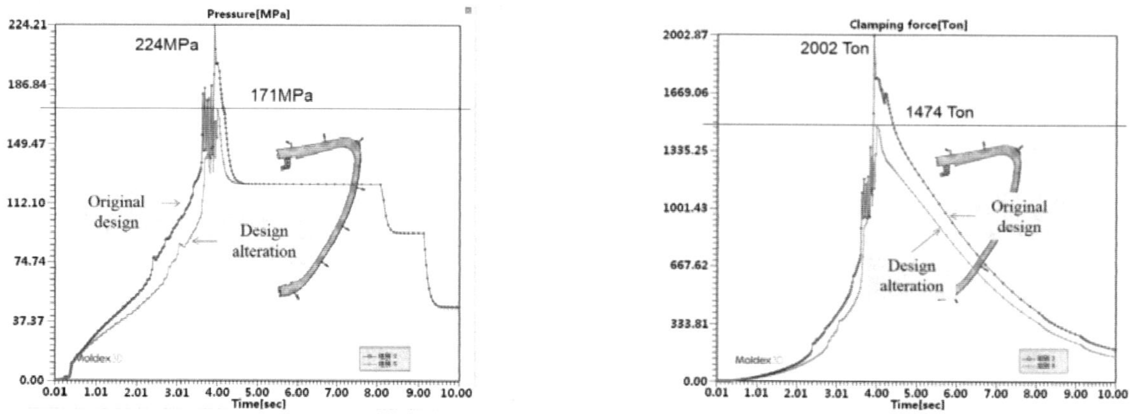

(a) Pressure curve of rubber inlet and (b) Curve of clamping force

Figure 9. Contrast curve

By comparing the warpage displacement of the two schemes with Fig.10, the result is enlarged by 5 times. It is obvious that the warpage displacement of the modified design is smaller than that of the original scheme. In the final maximum deformation, the optimization ratio is 24% from 18.5 mm to 14.1 mm.

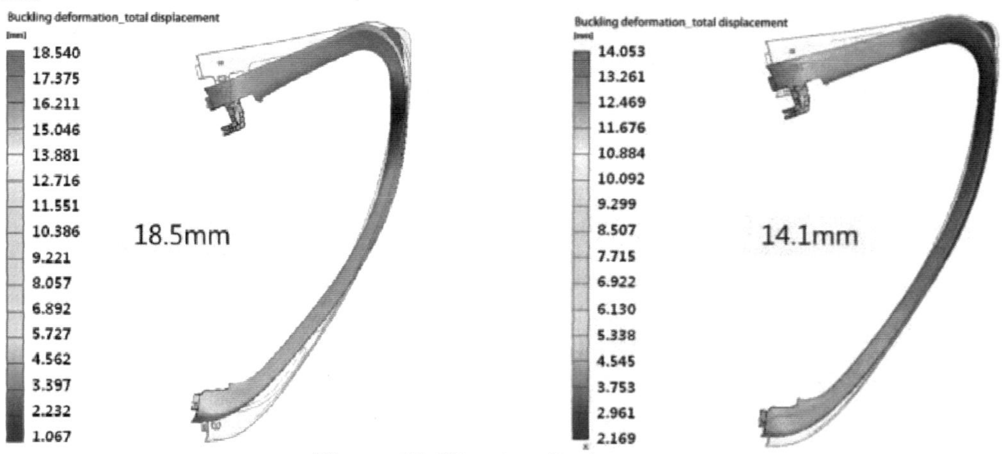

Figure 10. Warping displacement

4. Physical verification

Based on the above CAE flow analysis results, the scheme after the change of gate position is handed over to Swift Company for the actual test model comparison. The field test model and the forming product are shown in Fig.11. From the drawing of the forming product, it can be seen that the warping displacement of the right figure is smaller than that of the left one, which is close to the result of Moldex3D analysis, and the warping displacement has been improved obviously.

(a)Field trial

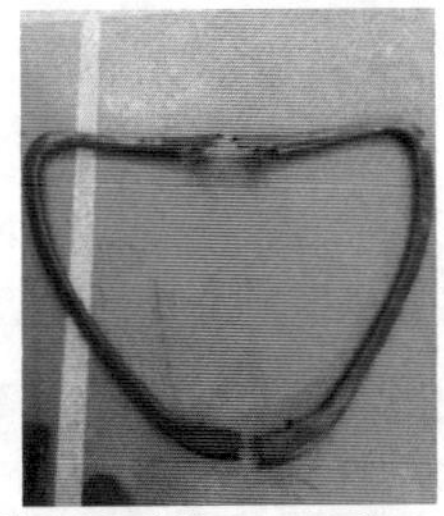
(b) Molding products

Figure 11. Field trial and molding products

5. Summary

Based on CAE technology, the optimization of gate position for warpage displacement is analyzed by Moldex3D flow analysis software. Through the comparison of the results, it is found that the warpage displacement is reduced and the forming quality of the product is improved by changing the gate position of the design compared with the original design. Through analysis and comparison, it is verified that the modified design has better analysis results than the original one：

The forming pressure is reduced from 224 MPa to 171 MPa (single cavity). At the same filling speed, the filling pressure is smaller.

The clamping force decreases from 2003 Ton to 1474Ton (single die hole), and the requirement of clamping force decreases with the decrease of forming pressure.

The warpage displacement was reduced from 18.5 mm to 14.1 mm by adjusting the gate position through changing the design, and the overall optimization was 24%.

However, with the continuous progress of science and technology, computer CAE simulation is more and more suitable for actual production, which makes it a stable reference before the actual production of finished products, and greatly improves the quality of product shaping and reduces the waste of resources.

6. Reference

[1] Ma, S.H. Forming technology of the plastic weather strips used in the bus doors and windows Engineering Plastics Application (04) 1989:PP.22-26.

[2] Papathanasiou TD, Guell DC. Flow-induced alignment in composite materials Cambridge: Woodhead 1997:pp. 127–30.

[3] Prashant, T., Alok, S.C., A, S., Raj, K.E., R, P. Estimation of Shrinkage & Distortion in WaxInjection using Moldex3D Simulation Materials Today 5 2018: pp.19410 - 19417.

[4] Christian H., Julian H. Process control strategies for injection molding processes with changing raw material viscosity. Journal of Polymer Engineering 38(5) 2018:pp.483-492.

[5] I S Stepanenko, A I Khaimovich. GTE blade injection moulding modeling and verification of models during process approbation Materials Science and Engineering 177 2017:pp.1-7.

[6] Tzu-Chau C.,Chao-Tsai H.,Yan-Chen C.,Wei-Da W.,Cheng-Li H., Chen-Yang L., Lung-Wen K. Material saving and product quality improvement with the visualization of hot runner design in injection molding International Journal of Precision Engineering and Manufacturing 14(6) 2013:pp.1109-1112.

IOP Conf. Series: Materials Science and Engineering **627** (2019) 012009 doi:10.1088/1757-899X/627/1/012009

The Study of Geometric Parameters of the Mong Impact on Natural Frequency

Joompon Bamrungwong[1,*] and Bancha Arthibenyakul[2]

[1]Industrial Physics and Medical Instrumentation, Faculty of Applied Science, King Mongkut's University of Technology North Bangkok, 10800 Bangkok, Thailand
[2]Industrial Physics and Medical Instrumentation, Faculty of Applied Science, King Mongkut's University of Technology North Bangkok, 10800 Bangkok, Thailand

joompondej@gmail.com*

Abstract. The Mong or Tam-tam is sometimes also called Gong, which is a percussion instrument in Thailand. The Mong with boss generates a tone with less shim and a one-key sound. A player has to percuss the Mongs to create a rhythm with a hard rubber mallet. It was originally used for signaling peasant workers in from the fields and telling time in Thai because the Mongs are loud enough to be heard from up to 8 km away. The sound of the Mong is usually generated from the vibration of its own structure. The aims of this analysis were to study the eigen modes and identify the important parameters to generate a natural frequency, in which the boss of the Mong are excited. Finite element analysis and the design of the experiment were used to find the relation of geometric parameters that impact the eigen mode. In this paper, the inertia relief method was used to study the natural frequency of the Mong, while the spectrum analyzer was used to measure the eigen mode. The results showed that the diameter of the horizontal flange and the thickness of the Mong had a significant effect on the eigen natural frequency, which is similar to the Gong. This implies that the thickness and horizontal flange diameter must be increased if wanting to use the Mong with many key tones. The analysis agreed with the actual behavior of the structure of the Gong of Khong Wong Yai.

1. Introduction

The melodies and rhythms are necessary to every form of music in all cultures. A Thai percussion instrument that looks like a Khong (Ḳĥxng) is sometimes called Mong (H̄om̀ng) in Thai. In English, it is called Gong or Tam-Tam. A tam-tam is a flat-faced gong of indefinite pitch. In other words, it does not have a boss at its center. Gongs have a boss at the center or a small boss more around the center boss. Its first use may have been to signal peasant workers in from the fields and telling time in ancient Thailand because the Mongs are loud enough to be heard from up to 8 km away. Nowadays, it is used as a percussion instrument often involving some kind of striking on the object, which causes vibrations that produce the sound. The boss or nipple is embossed in the center of the Mong. The Mong is made of circular-shaped metal plates with the same thicknesses, which are struck with soft beaters. The Mong does not need to be tuned with beeswax under the circular metal disc like the Khong. Mongs are now used in Thai temples as well as Chinese temples for worship. They come in various sizes of approximately 15 to 51 cm [1]. Mongs have varying degrees of quality and resonance with a tone and less shim. However, the engineering principles were applied to design and produce handmade gongs, while they were constructed with the experience and skill of the maker. Thus, the application of finite element analysis was used to better understand the geometry of the Mong.

Bor-Tsuen Wang studied the Chinese Gong by finite element method and attempted to verify by experimental modal analysis (EMA) [2]. However, Euler-Bournoulli's equation is often used for studying the vibration of the object. Four parameters are needed for the solution of the equation for only fixed supports [3]. Actually, the Mong is suspended vertically by a cord passed through two holes adjacent to the top rim of the bowl structure, which can be assumed as free in the space, not fixed support. Thus, it is impossible to determine the eigen modes of the gong by Euler-Bournoulli beam theory [4,5]. Joompon B. and Pattaraweerin W. studied the factors of the Gong impacting the natural frequency by using inertia relief, which found that the main effect is the diameter. The thickness of the flange and the thickness of cylinder of the Gong are important factors for tuning the sound. Furthermore, this study found that it is impossible to make the Gong without beeswax. It always needs beeswax for tuning the sound [6]. This research proposed to investigate the structure and modes shape of the Mong with no constraint body using inertia relief by FEM. The results of finite element analysis were validated by comparison with the experiment results.

2. Materials

An individual Mong has the same shape, but some Mongs have different structures by adding smaller nipples to emboss around the center of the Mong. In this study, the structure of the Mong was analyzed regarding the generated natural frequency. Starting from a Mong with no boss (Tam-Tam), a Mong with one boss at the center of the Mong and a Mong with various boss around the center of the Mong were studied. Although the general shapes of both the Mongs are the same, their sizes are different. Therefore, the eigen modes were validated by spectrum analyzer. The 3D model with actual size 25 cm in diameter of the Mong was created for analysis by finite element software. A Tam-Tam is shown in Fig. 1, while a type of Mong is shown in Fig. 2. The Mong with single boss is embossed at the center of the Mong, as shown in Fig. 2a, while the Mong with added smaller boss is embossed around the boss center of the Mong, as shown in Fig. 2b
In this study, the Mong was made of brass (Basically used for the Thai Mong). The engineering data of material properties used for FEA are density 8,600 kg/m3, Young's modulus 1.06 105 MPa and Poisson's ratio 0.31.

Figure 1. The Tam-Tam

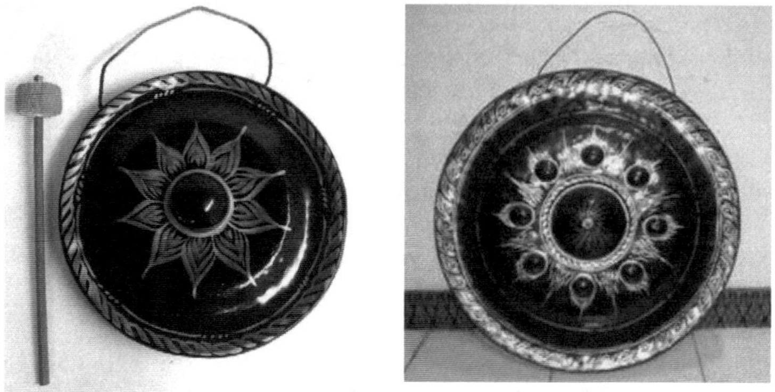

(a) (b)

Figure 2. (a) The Mong is 25 cm in diameter (b) The Mong with added smaller boss embossed around the bossed centre.

3. Method

Due to the vibration of the Mong being a free vibration, it can be considered as having a free elastic body without constraint. In previous research [4], the inertia relief approach was combined with finite element analysis in the modeling and analysis of the unconstrained systems. In this study, the commercial finite element package is used for analysis. The Mong with 250 mm diameter, 0.5 mm thickness and 40 mm vertical flange were studied. The Mong with no boss was the first study (Like a Tam-Tam), while the Mong with a single boss at the center was studied with the same size of the previous Mong analyzed. After that, the Mong with given a smaller boss embossed in around the bossed center of the Mong. In addition, the name of the parameters for this study is shown in Fig.3a. Fig.3b shows the parameters that were studied, with A being the vertical flange, B being the angle of the vertical flange, C being the thickness of the vertical flange, D being the diameter of the horizontal flange, and E being the diameter of the boss.

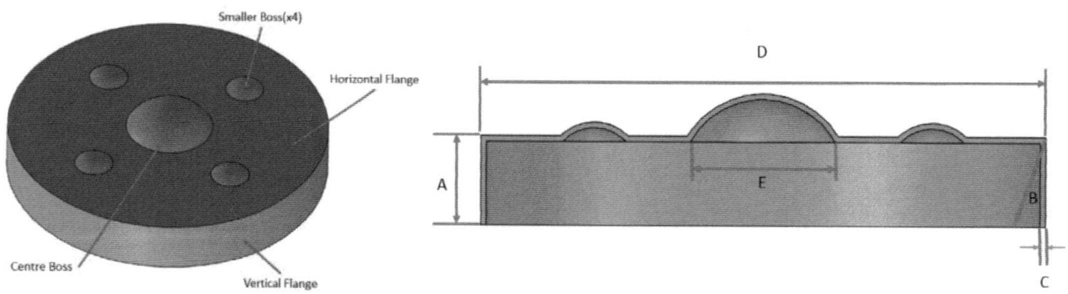

(a) (b)

Figure 3. The name of parameters (a) and the parameters studied (b)

The frequencies in hertz were recorded to study the natural frequency of the Mong. The Mong was considered as having no constraint on the Mong. For the sake of simplicity, the beam was assumed as two masses, as shown in Fig. 4.

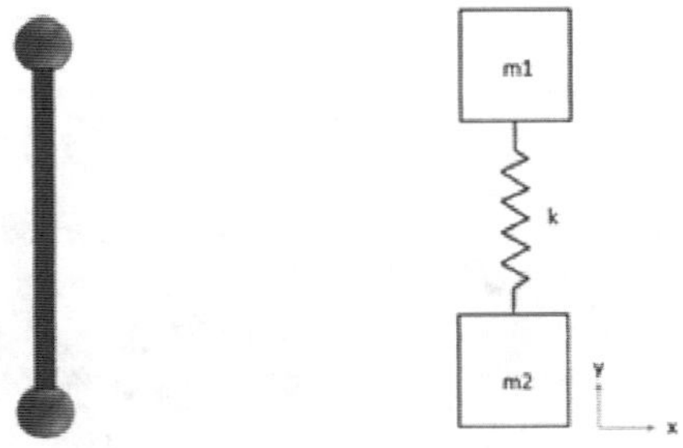

(a) (b)

Figure 4. (a) The beam was assumed as two masses

(b) The beam was simplified as two masses and a spring

The two masses (m1 and m2) were connected by a spring having stiffness k and there was no external force applied. Assume that both masses have the same weight, in which W1 and W2 are the weight of m1 and m2, respectively. Assume the spring and two masses can only move vertically (y direction in Figure 5). Because of the applied force, the system will move freely in a vertical direction only. The mass m1 starts moving downwards when compressing the spring, while the mass m2 is at the rest while also compressing the spring. When the spring is compressed enough to produce a reaction force more than W1+W2, m2 will start moving downwards. After some time, the system will arrive at a steady state, in which both m1 and m2 are moving in constant acceleration together. Considering the overall spring mass system, the equation of motion for free vibration can be written as:

$$a\begin{bmatrix} m_1 & 0 \\ 0 & m_2 \end{bmatrix} + \begin{bmatrix} k & -k \\ -k & k \end{bmatrix} \begin{Bmatrix} x_1 \\ x_2 \end{Bmatrix} = \begin{Bmatrix} 0 \\ 0 \end{Bmatrix} \tag{1}$$

Theoretically, it is impossible to determine the position of the two masses because they are not constrained in the system. Nevertheless, the relative displacement can be found by restraining either x1 or x2.

If $x_1 = 0$, then $x_2 = \dfrac{am_1}{k}$

If $x_2 = 0$, then $x_2 = \dfrac{am_2}{k}$

This shows that the relative displacements x1 – x2 were the same for either constraint. The result of the acceleration and the deformation of systems were not affected by the selection of constraints in this system, which is called the inertia relief analysis [5]. Most frequently, the inertia relief approach is combined with finite element analysis in the modeling and analysis of unconstrained systems.

4. Results

Five cases of the Mong model have been studied by FEA simulation. Only the experiment of case 4 has been done. The result from the experiment and FEA result were compared, as shown in Fig. 5.

IOP Conf. Series: Materials Science and Engineering **627** (2019) 012009 doi:10.1088/1757-899X/627/1/012009

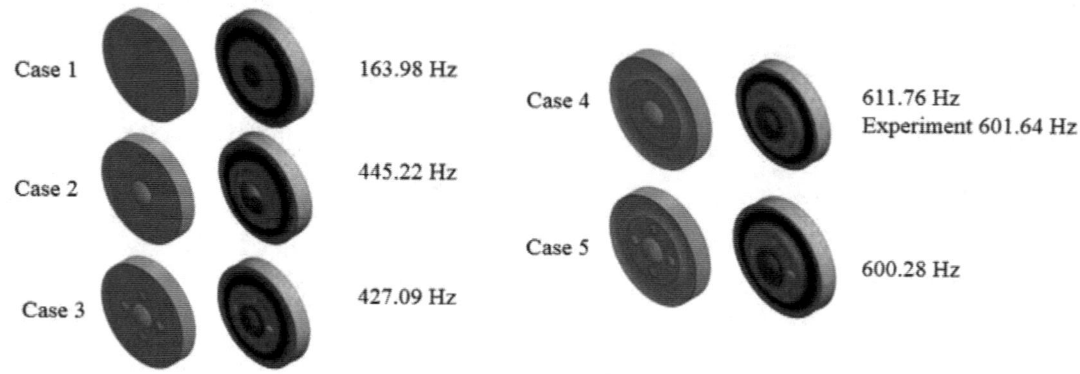

Figure 5. The same order eigen mode resulting from FEA and the experiment

The first case is the Mong with no boss at the center. Case 2 is the Mong with one boss at the center. Case 3 is the Mong with one boss at the center and a smaller boss around the center boss. Case 4 is the Mong with one boss at the center and a curved horizontal flange. The last case is the Mong with one boss at the center and a curved horizontal flange as well as a smaller boss around boss center. The parameters in Fig. 3b were varied by more than 3 points, as shown in Fig. 6 to Fig. 8. From the analysis results, the parameters severe by consider the slope of the graph, shown parameters A, B, C, D and E in Table 1 and found that parameter C had the most significant effect on natural frequency by consideration from the value of the unit per effect (slope).

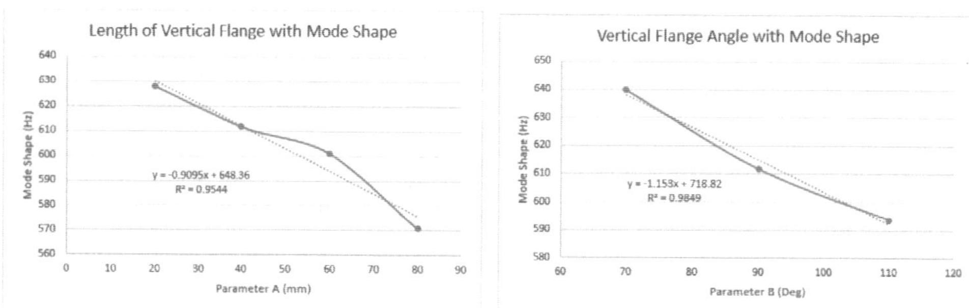

(a) (b)

Figure 6. (a) Length of the vertical flange and mode shape
(b) Angle of the vertical flange and mode shape

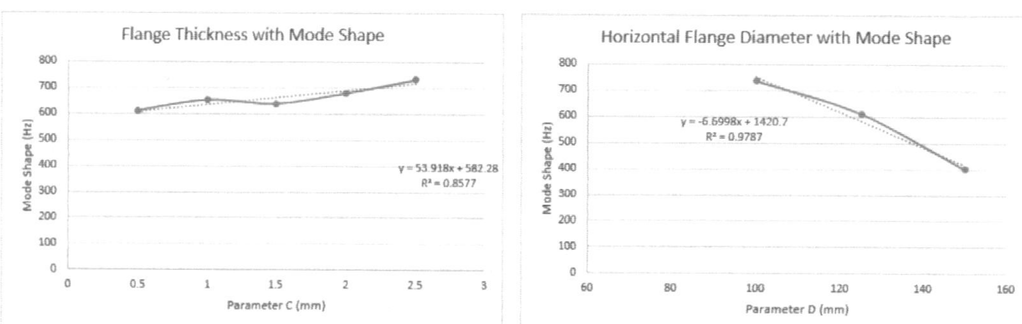

(a) (b)

Figure 7. (a) Thickness of the flange and mode shape
(b) Diameter of the horizontal flange and mode shape

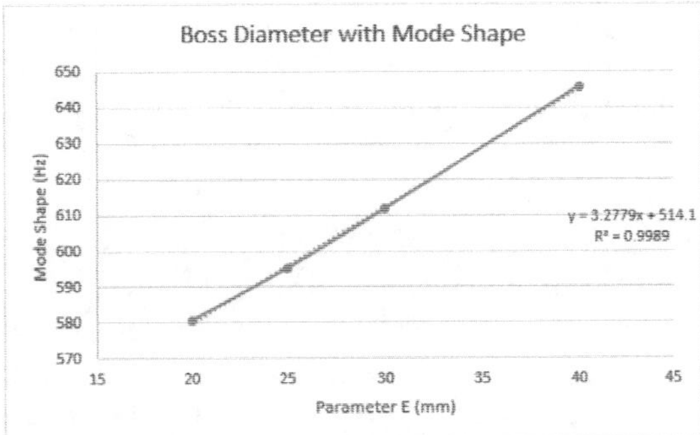

Figure 8. Diameter of the boss and mode shape

Table 1. Unit per effect of the Mong's parameters

Parameters	Remark	Unit	Unit/Effects
A	Length of Vertical Flange	mm	1
B	Vertical Flange Angle	Deg	1.2
C	Flange Thickness	mm	53.9
D	Horizontal Flange Diameter	mm	6.7
E	Boss Diameter	mm	3.3

The design of experiment (DOE) was created to find the parameter relationship. Parameters A, B, C, D and E were used to create the run in design of experiment, as shown in Table 2. The finite element analysis was used to study the natural frequency following the run order.

Table 2. Created Factorial Design for Run

StdOrder	RunOrder	CenterPt	Blocks	A	B	C	D	E	Freq. (Hz)
8	1	1	1	60	110	2	100	25	874.16
31	2	1	1	20	110	2	150	40	494.76
23	3	1	1	20	110	2	100	40	962.94
19	4	1	1	20	110	0.5	100	40	1030.2
26	5	1	1	60	70	0.5	150	40	476.91
2	6	1	1	60	70	0.5	100	25	1127.5
32	7	1	1	60	110	2	150	40	453.01
27	8	1	1	20	110	0.5	150	40	478.98
9	9	1	1	20	70	0.5	150	25	496.57
13	10	1	1	20	70	2	150	25	409.85
3	11	1	1	20	110	0.5	100	25	1051.8
29	12	1	1	20	70	2	150	40	470.66
6	13	1	1	60	70	2	100	25	1078.2
10	14	1	1	60	70	0.5	150	25	490.81
33	15	0	1	40	90	1.25	125	32.5	709.9
17	16	1	1	20	70	0.5	100	40	1132.8
15	17	1	1	20	110	2	150	25	441.09
16	18	1	1	60	110	2	150	25	415.78
24	19	1	1	60	110	2	100	40	845.19
28	20	1	1	60	110	0.5	150	40	481.54
18	21	1	1	60	70	0.5	100	40	1125.2
12	22	1	1	60	110	0.5	150	25	441.57
30	23	1	1	60	70	2	150	40	438.7
5	24	1	1	20	70	2	100	25	1175.6
20	25	1	1	60	110	0.5	100	40	987.8
4	26	1	1	60	110	0.5	100	25	1025
21	27	1	1	20	70	2	100	40	1174.7
1	28	1	1	20	70	0.5	100	25	1133.6
7	29	1	1	20	110	2	100	25	987.74
14	30	1	1	60	70	2	150	25	388.47
11	31	1	1	20	110	0.5	150	25	453.51
25	32	1	1	20	70	0.5	150	40	476.07
22	33	1	1	60	70	2	100	40	1059.7

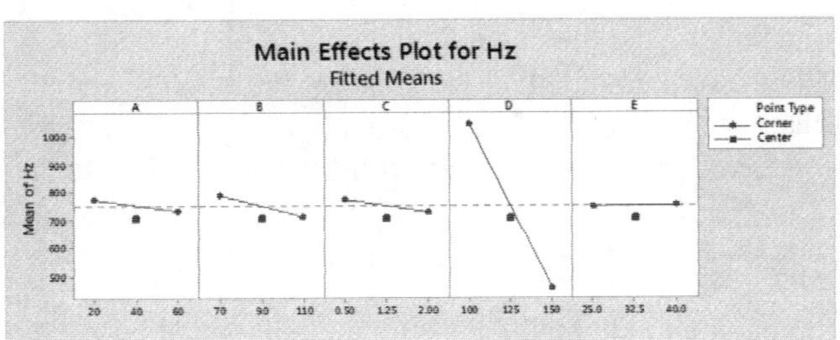

(a)

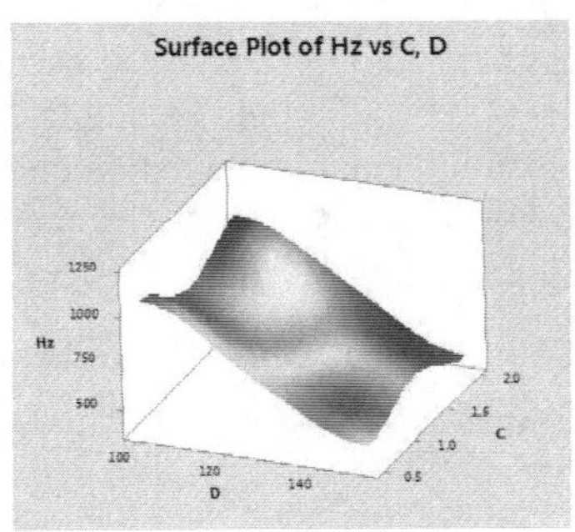

(b)

Figure 9. (a) The main effects plot of the five factors and (b) the surface plot

The results show that factor D was the first parameter impacting the natural frequency, as shown in Fig. 9(a). P-Value less than 0.05 when considering R-Sq (adj) was found to be high enough at 99.36%. The preliminary discussion found that factors C and D impacted the eigen mode. Thus, the surface plot of the natural frequency and factors C and D is shown in Fig. 9b.

5. Conclusion and Summary

This study revealed that the Mong without a center boss has a lower tone key than the Mong with center boss. This can be considered that they are Tam-tam and Mong, respectively. The Mong with added curve on the horizontal flange has a higher tone key than a flat horizontal flange. Adding a smaller boss around the center boss causes a decrease in the natural frequency. The initial and designed experimental study found that the thickness and diameter of the horizontal flange of the Mong was a key parameter related to tuning the natural frequency, which is in agreement with the actual percussion instrument, like the Thai Gong (Khong Wong Yai). The response surface function plot reveals that the smaller diameter of the horizontal flange of the Mong has a higher tone key than a horizontal flange with a larger diameter. If a high tone key is required, a horizontal flange with a small diameter and low thickness should be used. However, the thickness of the Mong has to be optimized if needed. The natural frequency of the Mong has the eigen modes generated from bending mode, which involves the entire horizontal flange of the boss moving upwards and downwards. It should be noted that drilling holes to hold the gong should be near a horizontal flange to balance the Mong in a vertical direction.

6. Reference

[1] Gong. [Online]. 30 May 2016, available: https://en.wikipedia.org/wiki/Gong.
[2] Bor-Tsuen Wang,2011, Intergation of FEA and EMA Techniques for Percussion Instrument Design Analysis. International Conference on System Science and Engineering, China (2011).
[3] J. Bamrungwong, 2006, Analysis and Design The Natural frequency of a Treble Gamelan Using FEA. MENETT#20, Thailand (2006).
[4] J. Bamrungwong ,2016, The Natural frequency Analysis of the Khong Wong Yai by FEM. International Conference Facilitating Autonomous Learning via Research-Based Approaches (FCAL), Thailand (2016).
[5] J. Bamrungwong, 2016, The Application of Inertia Relief Method in Ranat and Xylophone, International Conference on Electrical, Mechanical and Industrial Engineering, Thailand (2016)
[6] J. Bamrungwong and Pattaraweerin W., The Effect of the Khong Wong Yai Parameters on Sounds by FEM, International Conference on Advanced Materials and Engineering Materials, March 11-12nd, Singapore, 2017.
[7] J. Bamrungwong and W. Amarin, 2017, DOE for Study of the Sound Factor of the Gong, International Conference on Industrial Engineering, Thailand (2017).
[8] Raymond A. Serway and John W. Jewett, Jr. “Physics for Scientists and Engineers with Modern Physics” 9th ed, Cengage Learning, 2014
[9] Benson H. Tongue, “Principles of Vibration” 1st ed, Oxford University Press,1996.
[10] Daryl L. Logan, “A First Course in the Finite Element Method” 4th ed, Cengage Learning, 2007
[11] Robert D. Cook, “Finite Element Modeling for Stress Analysis” John Wiley and Son Inc., 1995
[12] Bathe, “Finite Element Procedure”, Prentice Hall of India Private Limited,2007

IOP Conf. Series: Materials Science and Engineering **627** (2019) 012010 doi:10.1088/1757-899X/627/1/012010

Research on the Trend of Structural Failures from Vehicle Body Spreader Based on ANSYS Analyses

Duo Sun[1], Tianmin Guan[1] and Lei Lei[1]

[1] School of Mechanical Engineering, Dalian Jiaotong University, Dalian 116028, Liaoning, China Type the author addresses here
[a]sddjtu@163.com

Abstract. The operating conditions of the vehicle body spreader have been analyzed in this paper, and all the typical conditions have been applied with the nonlinear finite element analysis of the contact, aiming to ensure the safe utilization of the body spreader of tooling equipment for the production of rail vehicles. The potentially risky component locations are predicted and the weak spots are discovered on the basis of the analysis results of the static strength. The anticipation of the component failure trends shows a guiding role in the failure prediction in actual production and facilitates accident prevention.

1. Introduction

Vehicle body spreader is the process and construction installation equipment of Changchun Railway Vehicle Co., Ltd., and this equipment is used for hoisting the vehicle body. The vehicle body spreader is usually used by pairs, two overhead travelling cranes are used to hoist the construction installation equipment and vehicle body[1]. Conduct finite element analyses on the vehicle body spreader in order to guarantee the equipment to be safe, scientific and rational. Make prejudgment on the trend of part failures through simulating the static strength and rigidity calculations under multiple working loads[2], to find out the weak points, and therefore to propose bases to maintenance and to prevent from accidents.

2. Contact Nonlinear Finite Element Analysis Technology

ANSYS finite element software is a multi-purpose finite element calculation program, which could be used for seeking solutions to issues as structures, fluid, electric, electromagnetic field and collisions, etc. ANSYS supports three contact modes: point-point, point-surface and surface-surface, and the contact unit used for each contact mode is applicable for a certain issues[3]. In order to create the models for contact issues, first of all, it must be realized that which parts in the model might contact each other, and the corresponding component of the model shall be a node if one of the two in mutual contact is a point. The corresponding component of the model shall be a element if one of the two in mutual contact is a surface[4]. For example, as for the beam element, shell element or solid element, the finite element will recognize the possible contact pairs by means of the designated contact elements.

3. Establishment of Vehicle Body Spreader Finite Element Model

PRO/E 3D mapping software is adopted for 3D modeling of the vehicle body spreader, and HyperMesh v10.0 is used for finite element modeling of the equipment, and the large general finite element analysis software ANSYS is used for static strength analysis of the vehicle body spreader. Perform 3D modeling on the vehicle body spreader according to the drawing data of vehicle body spreader and the relationship drawings between parts. The 3D model of vehicle body spreader is as indicated by Fig. 1.

Then guide the 3D model into HyperMesh for finite element meshing, and establish the finite element model. All the structures contributing to overall rigidity and partial strength of the vehicle body spreader framework shall be considered. In order to achieve the accuracy of calculations, the finite element model of vehicle body spreader equipment mainly consists of hex solid elements[5]. The finite element model of the vehicle body spreader is as indicated by Fig. 2.

The defining of contact pairs shall be completed after completing the finite element modeling for all parts. There are two selecting methods among the selections of surface-surface contact pairs, one is selecting solid-to-solid; the other is producing two shells on the surfaces of two contact bodies, and defining the contact relationship for the two shell element surfaces. As for the parts of the equipment are large, many calculations will be increased if the solid contact is selected, so defining the contact relationship for the surface shell elements is selected is more rational for calculations, in order to facilitate calculations. The contact surface defined is as indicated by Fig. 3.

a.Overall assembly drawing b. Drawing of part

Fig. 1 3D Model of Vehicle Body Spreader

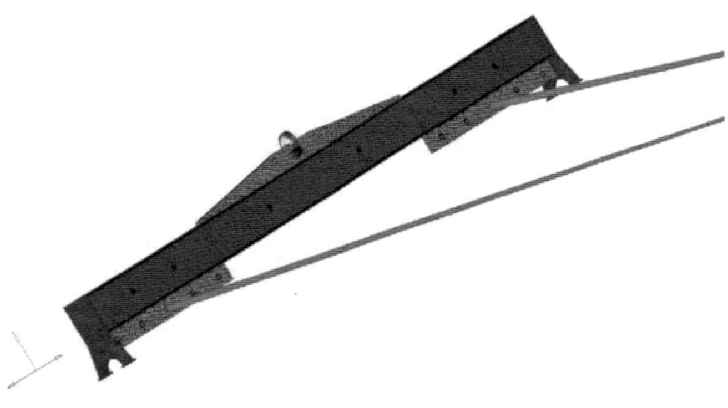

Fig. 2 Overall view to finite element model of the vehicle body spreader

Fig. 3 Diagram of the Contact Surface Defined

4. Loading Conditions and Boundary Conditions

The working conditions for calculations could be classified into 12 according to the working spaces in actual use and according to different working spaces and the loading conditions, and they are respectively the outside, middle and inside general loads and overloading conditions of the simulated frocks and working pieces in the process of operation, the outside, middle and inside eccentric overloading conditions and the overloading conditions with outside high-low, middle high-low and inside high-low. Refer to Table 1 for details of the loading conditions and boundary conditions[6].

Table 1. Detailed Table for Partial Loading Conditions of Vehicle Body Spreader

No.	Loading Conditions	Marks and Explanations		Loading Values	Load Applying Positions
1	External rated loading condition	FY	Each vehicle body spreader	12t	Center of Vehicle Body
2	Middle rated loading condition	FY	Each vehicle body spreader	12t	Center of Vehicle Body
3	Internal rated loading condition	FY	Each vehicle body spreader	12t	Center of Working Piece
4	External eccentric overloading condition	FY	Each vehicle body spreader	12t	Center of Vehicle Body

5. Calculation Results of Static Strength

Footnotes should be avoided whenever possible. If required they should be used only for brief notes that do not fit conveniently into the text.

By means of the ANSYS post-processing treatment function, the calculation results of each working condition could be read, hence, the cloud charts for strains and stresses of various working conditions could be obtained, and the position of secondary maximum stress point could be sought through deducting the maximum stress point and its stress value could be displayed[7]. 5 stress points are selected in turn in order to find the positions with possible hazards as much as possible.

External Rated Loading Conditions

Under function from the external rated loading conditions, the cloud charts of strain and Von. Mises stress for the vehicle body spreader is as indicated by Fig. 5. The maximum Von. Mises stress of the vehicle body spreader is 134MPa, which occurs at the position of the upper lifting hook shaft; the secondary maximum Von. Mises stress is 84MPa, which occurs at the position of the connecting shaft between the most outside mounting plate and transom; the third maximum Von. Mises stress is 82MPa, which occurs at the position of the most outside lifting hoop hole; the fourth maximum Von. Mises stress is 81MPa, which occurs at the position of the connecting shaft between the middle mounting plate and the transom; the fifth maximum Von. Mises stress is 33MPa, which occurs at the position of the second outside lifting loop hole.

IOP Conf. Series: Materials Science and Engineering **627** (2019) 012010 doi:10.1088/1757-899X/627/1/012010

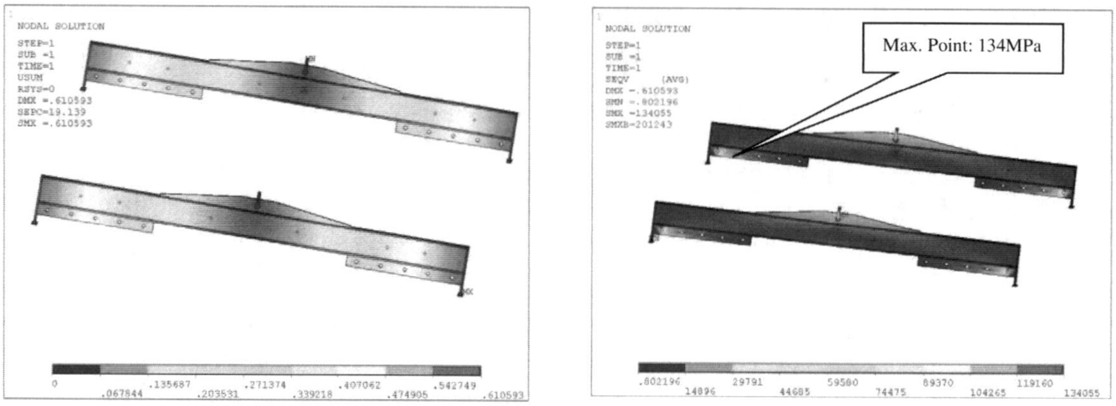

a. Cloud Chart of Deformation b. Cloud Chart of Stress

Fig. 4 Cloud Charts for Deformation and Stress of Vehicle Body Spreader under the External Rated Loading Conditions

Middle Rated Loading Conditions

Under function from the middle rated loading conditions, the cloud charts of strain and Von. Mises stress for the vehicle body spreader is as indicated by Fig. 6. The maximum Von.Mises stress of the vehicle body spreader is 136MPa, which occurs at the position of the upper lifting ring shaft; the secondary maximum Von.Mises stress is 52MPa, which occurs at the position of the middle lifting ring hole; the third maximum Von. Mises stress is 51MPa, which occurs at the position of the connecting shaft between the most outside mounting plate and the transom; the fourth maximum Von. Mises stress is 19MPa, which occurs at the position of the inside second lifting ring hole; the fifth maximum Von. Mises stress is 13MPa, which occurs at the position of the lifting ring mounting plate's corner.

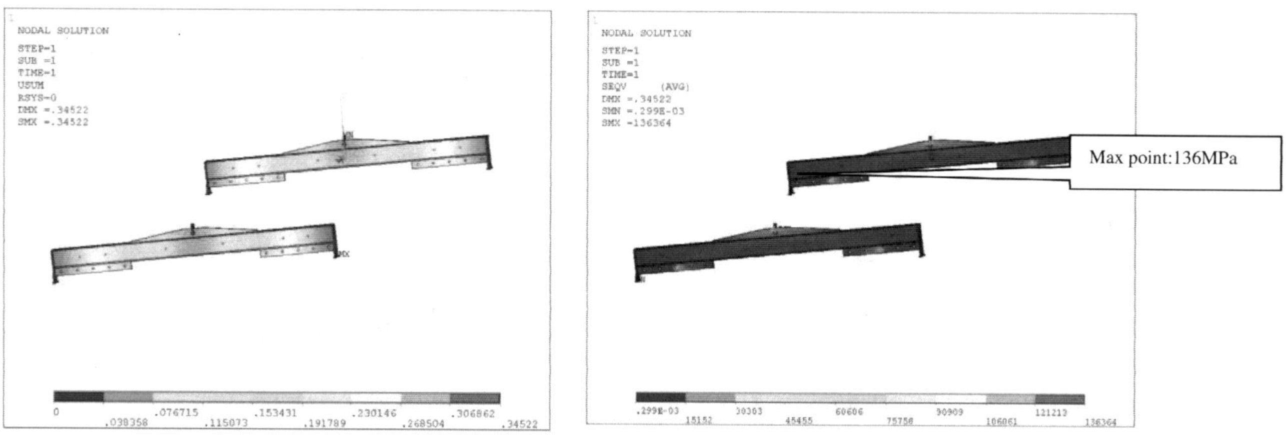

a. Cloud Chart of Deformation b. Cloud Chart of Stress

Fig. 5 Cloud Charts for Deformation and Stress of Vehicle Body Spreader under the Middle Rated Loading Conditions

Analyses on Calculation Results

The summary of calculation results for static strength of vehicle body spreader is as indicated by Table 2.

Table 2 Summary of Calculation Results for Nonlinear Static Strength of Vehicle Body Spreader

Working Conditions for Calculations	Maximum Deformation (mm)	Von. Mises Stress (MPa)				
		Maximum Stress	Secondary Maximum Stress 2	Secondary Maximum Stress 3	Secondary Maximum Stress 4	Secondary Maximum Stress 5
Condition 1	0.610	134	84	82	81	33
Condition 2	0.676	138	93	86	82	32
Condition 3	0.345	136	52	51	19	13
Condition 4	0.413	163	62	53	23	15
Condition 5	0.194	116	89	78	77	28
Condition 6	0.232	139	138	79	77	34

As indicated by the results of Table 2, working condition 10 is the most hazardous condition, and its deformations and stresses from various parts are the maximum values of the various working conditions, and this working condition shall be avoided as much as possible in use of the equipment.
The maximum stress of the equipment is located at the contact position between the pin hole and the lifting hook hinge pin, so the surfaces of the pin hole and lifting hook hinge pin shall be critically monitored. Large stresses occur at the welding position between the lifting bushing and the mounting plate, and 100%UT flaw detection treatment shall be conducted before ex-factory.
The major force bearing points occur at the welding positions between the primary mounting plate and the end of U-steel rib plate. Considering from the comprehensive perspective of strain and stress, the pin hole and lifting hook hinge pin are the trend parts for equipment failures and damages; secondly is the welding position between the lifting bushing and the mounting plate.
As indicated by the analysis cloud charts of strain and stress, evidently there is high stress and deformation existing at the external side of the welding position between the U-steel rib plate and the girder rib plate, and shall be considered as the critical object for maintenance.
Refer to Table 3 for summary of the critical monitoring parts for vehicle body spreader.

Table 3 Summary of Critical Monitoring Parts for Vehicle Body Spreader

No.	Critical Monitoring Parts
1	The position of hoisting ring shaft
2	The hoisting ring position
3	The position of shaft connecting the mounting plate and transom
4	The position of hoisting ring hole
5	The position of corner to the hoisting ring mounting plate

6. Conclusions

In order to guarantee the vehicle body spreader to be safe, scientific and rational, the application working conditions for equipment are selected in the paper, contact nonlinear finite element analyses are conducted on various typical working conditions. It's innovatively proposed in the paper that obtaining the positions of the parts might arise with hazards and the contact positions between the pin hole and lifting hook hinge pin through analyzing the results of static strength, and it's the same with the situation of the actual failures. Make prejudgment on the trend of part failure through finding out the weak points. It will guide the prejudgment of failures in actual production and will prevent from accidents. Meanwhile, it will make preparations for the establishing of equipment reliability analysis database in the next step through making prejudgment on the trend of part failures from the equipment.

7. References

[1] Cai X, Sixin Q I, Chunlai W U, et al. Investigation on the Protect Criteria of Aero-engine Static Strength Design[J]. Journal of Mechanical Engineering, 2017.

[2] Makhutov N A. Strength, service life, and safety of machine systems[J]. Journal of Machinery Manufacture & Reliability, 2014, 43(3):217-232.

[3] Affolter C, Piskoty G, Koller R E, et al. Limitations of analytical strength verifications with local effects and nonlinearities: A case study on a failed high rack rail[J]. Engineering Failure Analysis, 2015, 56:28-38.

[4] Yaohui L U, Xiang P L, Xie N, et al. FATIGUE STRENGTH ANALYSIS OF CONNECTING ROD BOLT FOR LARGE POWER DIESEL ENGINE[J]. Journal of Mechanical Strength, 2016.

[5] Qingwu Wang, Fang Zuo, Renxi Hu, et al. Examples for Applications of ANSYS 10.0 Machinery Design (Version 2) [M]. Mechanical Industry Press, January 2006

[6] Hongyun Li, Shexu Zhao, Yan Sun. ANSYS 10.0 Bases and Engineering Applications [M]. Mechanical Industry Press, 2008

[7] Ke'an Li, Sheng Zhou and Rongfan Cui, et al. Analysis on Structures of Finite Element for Heavy-Duty Gas Turbine Spreader [J]. Journal of Hunan College, March 2008

Upgrading Strategy of Small and Medium Manufacturing Enterprises (SMMEs) to Smart Manufacturing

Baolei Zhang [1,2], Fuquan Zhao[1,2] and Zongwei Liu[1,2]

[1]Tsinghua Automotive Strategy Research Institute, 100084 Beijing, China
[2]State Key Laboratory of Automotive Safety and Energy, 100084 Beijing, China

liuzongwei@tsinghua.edu.cn

Abstract. The contribution of Medium Manufacturing Enterprises (SMMEs) will not only weaken but also be strengthened under the background of smart manufacturing. It is the key point of achieving mass individualized production that lots of SMMEs with specialized and excellent manufacturing capability connect to smart manufacturing network. Nowadays, SMMEs of China are facing survival pressure and various severe challenge. Meanwhile, the developing experience and characteristics of SMMEs of China leads to the particularity of their upgrading process to smart manufacturing. The problems of process normalization, essential data, talents, and funds restrict SMMEs from upgrading to smart manufacturing seriously. It is very important for government, professional organizations, scientific research institutions, enterprises to do well on division of work and cooperation aiming at the key problems that enterprises are confronting now, in order to push forward the upgrading task to smart manufacturing together.

1. Introduction

With a new round of scientific and technological revolution profoundly changing the holistic society, traditional manufacturing industry is evolving into a new form under the impacts of new technologies such as Internet of Things, big data, cloud computing and 3D printing [1-3], necessitating changes of Medium Manufacturing Enterprises (SMMEs) [4-6]. Facing the future-oriented manufacturing industry, countries around the world have proposed their own development strategies. Based on a good industrial base, Germany has proposed the Industry 4.0 strategy, which has aroused strong repercussions worldwide. In order to revive the manufacturing industry, the United States put forward the development goal of the industrial internet in combination with its own information technology advantages. Japan has fully utilized its advantages in the field of robotics and proposed a robot-driven industrial revolution strategy. China has also clearly stated that it is necessary to speed up the construction of a manufacturing power. In terms of these manufacturing upgrading strategies, the major industrial countries are all oriented towards the full application of advanced information technology in the manufacturing industry, striving to achieve interoperability between production factors and integrating intelligent software and hardware into the whole process of production and management to finally achieve smart manufacturing [7-9]. Essentially, mass customization is the core goal and development trend of smart manufacturing [10-12].

It is necessary to pay special attention to the fact that the role of SMMEs will be further strengthened instead of being weakened under this new wave of industrial transformation and upgrading [13,14]. The integration of many high-quality small and medium enterprises with distinct manufacturing technologies and experience into the smart manufacturing system is the key to achieving mass customization. However, compared with the urgent need for smart manufacturing upgrades, China's SMMEs are still far behind in terms of the overall development level and facing a series of complex problems. How small and medium-sized manufacturing enterprises realize the upgrading and development of smart manufacturing has become one of the important issues to be solved immediately to build China into a manufacturing power. To this end, this paper systematically sorts out the current situation of China's SMMEs, and comprehensively studies the upgrading goals, major difficulties and effective measures of SMMEs in China for smart manufacturing, and present direction suggestions and policy reference for SMMEs to implement upgrading of smart manufacturing.

2. Development status and upgrading requirements of China's SMMEs

2.1. Development Status of China's SMMEs

2.1.1. Contributions of China's SMMEs. In the 1990s, developed countries shifted manufacturing industries to the outside world due to cost and other reasons, and China became the largest recipient of manufacturing capacity. The booming development of the manufacturing industry enabled China to quickly become a "world factory" and surpass the United States becoming the largest manufacturing power of the world in 2010. Thousands of small and medium enterprises serve as the basis for China's huge manufacturing system, which just like the thousands of capillaries of a huge living body that transport blood to such a "giant" as "Made in China". Since the end of the last century, small and medium manufacturing companies have provided more than 70% of new industrial output value, more than 60% of total exports, and more than 80% of urban jobs in China.

2.1.2. Classification of China's Small and Medium Enterprises. According to China's current classification standard for small and medium enterprises, enterprises in the industrial sector with employees less than 1,000 units or operation revenue less than 400 million CNY are defined as small and medium enterprises. Specifically, enterprises with more than 300 but less than 1,000 employees and operation revenue of 20 million CNY and above are medium enterprises; enterprises with more than 20 but less than 300 employees and operation revenue of more than 3 million CNY are small businesses. By the end of 2016, the total number of small and medium enterprises in China exceeded 24 million.

2.1.3. Characteristics of China's SMMEs. Some of China's SMMEs are to make up for the early market vacancies of large enterprises, and more are to provide support for large enterprises. Since most of them have been at the medium-and low-end of the industrial value chain since their establishment, they are generally short of core technologies and weak in independent innovation. With low added value, their products are less competitive than international rivals.

2.1.4. Survival Condition of China's SMMEs. In recent years, the continuous rise in labor costs has gradually deprived China's manufacturing industry of its traditional advantage of low cost. In 2003, the comprehensive labor costs of China were only about 40% of that of the United States. According to Oxford, the labor costs of China's manufacturing industry had approached that of the United States by the beginning of 2016. Meanwhile, the United States significantly reduced corporate income tax and strongly promoted "manufacturing returns"; while Southeast Asia and South Asia are taking advantage of the low cost of labor to absorb manufacturing capacity shifts. Therefore, China's manufacturing industry is facing severe challenges from developed and other developing countries. Specifically, SMMEs are under marked impacts.

On the other hand, China fails to pay enough attention to environmental cost management for economic development for a long time, which has caused environmental problems to intensify. China is gradually approaching the limits of natural resources carrying capacity. Therefore, there has been no living space for the previous production model with high emissions and high pollution. Increasingly stringent emission standards have also brought unprecedented pressure on companies, especially small and medium manufacturing companies.

2.2. Upgrade Requirements of China's SMMEs towards Smart Manufacturing

2.2.1. Upgrade of the Manufacturing Industries in Developed Countries. After the financial crisis in 2008, major developed countries re-recognized the importance of the industry-based real economy and launched national strategies to revitalize manufacturing industries. These industrial development plans of developed countries are not to continue the traditional industrial road, but aiming at a new round of scientific and technological revolution. The planes are committed to deeply integrating information technology and traditional industrial technology [15-18], to building cloud factories [19-21], complementing the shortcomings of the workforce via large-scale robotic applications [22], and further consolidating and strengthening industrial superiority through continuous innovation. Its focus lies in developing high-end manufacturing industry, creating high value-added links and restructuring the manufacturing industry chain [23].

In developed countries, particularly Germany and Japan, a large number of small and medium enterprises that are in absolute dominant positions in various sub-sectors have taken shape due to long-term industrial accumulation, namely "invisible champions" in sub-sectors. In the future, in the smart manufacturing system, these SMMEs are expected to form an interconnected manufacturing platform with large manufacturing enterprises, thereby greatly improving the overall competitiveness of the manufacturing industry [24]. In contrast, China's small and medium manufacturing companies are relatively weak, which has become one of the major bottlenecks hindering the upgrading of China's manufacturing industry towards smart manufacturing. To this end, how China's SMMEs can grasp strategic opportunities under current situations and realize transformation and upgrading towards smart manufacturing as soon as possible becomes crucial.

2.2.2. Upgrade Goals of China's SMMEs toward Smart Manufacturing. For upgrading towards smart manufacturing, China's SMMEs need to achieve two objectives from the internal to the external. The internal goal is to realize intelligent production, establish a complete corporate IoT, and have an efficient and integrated MES (manufacturing execution system) and SCADA (supervisory control and data acquisition) to realize the full digitization of business processes. The external goal is to gradually get integrated into the increasingly intelligent industrial chain and fully connected with upstream and downstream partners at the digital level, in order to achieve efficient interaction with other links of the industry.

3. Major difficulties faced by upgrade of China's SMMEs toward smart manufacturing

In terms of development from traditional manufacturing to smart manufacturing, China's SMMEs not only face a series of common basic problems, but also many specific difficulties.

3.1. Common Basic Problems Faced by Upgrade of China's SMMEs toward Smart Manufacturing

3.1.1 Lack of Correct Understandings of Upgrade Towards Smart Manufacturing. After the financial crisis in 2008, multiple factors such as continuous shrink of the market, labor constantly rise of costs and increasing pressure on environmental protection have significantly deteriorated the survival environment of substantial SMMEs, which places many operators in constant anxiety, who hence have a sense of urgency and show little interest in businesses that cannot immediately realize profitability. Although most small and medium manufacturing companies have realized that upgrading towards smart manufacturing is the future development direction of enterprises, they lack sufficient understanding of the urgency and long term, which results in slow actions and ineffective measures.

At the same time, many small and medium enterprises have serious deviations in understanding the relationship between smart manufacturing and lean production. Some companies believe that since smart manufacturing is an upgraded version of lean production, intelligent upgrade is enough while research and promotion of lean production are unnecessary, or upgrade towards smart manufacturing will not be considered before doing a good job in lean production. In fact, this is a completely wrong understanding. Smart manufacturing and lean production are two dimensions. The most important thing behind lean production is the concept of lean, which directly refers to the essential demands of the manufacturing industry and will never be outdated; smart manufacturing provides new technical means for a higher level of lean production [25]. In other words, lean production is the goal while smart manufacturing is the path, allowing no upgrade or replacement. SMMEs must take actions now to constantly improve their own hardware and software capabilities for smart manufacturing and try to do a better job in lean production in a down-to-earth manner [26].

3.1.2 Shortage of Talents. Most of China's SMMEs are mainly engaged in simple processing and manufacturing and employees are generally not highly educated, short of high-level professionals and information talents [27, 28]. In addition, in terms of upgrade towards smart manufacturing, SMMEs generally lack talents for planning, and even the on-the-spot management and maintenance personnel are seriously inadequate. Therefore, the shortage of talents is also one of the main factors hindering the upgrade of SMMEs towards smart manufacturing.

3.1.3 Low Level of Equipment. Many SMMEs have been at the low end of the industrial chain for a long time, resulting in low level of product technology, low accuracy of machining, outdated production equipment and low degree of automation [29, 30]. For smart manufacturing upgrades, priorities should be given to automation, digitization and connectivity of production equipment, which requires a substantial increase in investment, resulting in insufficient power of enterprises for upgrade.

3.1.4 Nonstandard Management. Due to low starting point, most China's SMMEs have been lacking standardized management processes since the foundation. In the process of upgrading towards smart manufacturing, it is difficult to digitize unsystematic management process, and even a successful digitization cannot truly meet the business demands. Therefore, the establishment of a scientific and standardized system management process is an important prerequisite for intelligent upgrade of SMMEs.

3.1.5 Financial Difficulties. Increasingly fierce market competition has caused continuous decline in profit of SMMEs in China, resulting in long-term lack of sufficient financial accumulation. Many SMMEs suffer from time-consuming collection period and get bogged in a financial strait. On the other hand, small and medium enterprises have difficulty in obtaining low-interest financial support from the financial sector due to the lack of sufficient physical collateral [31, 32]. Therefore, facing capital investment in the upgrade towards smart manufacturing, SMMEs lack adequate capabilities.

3.1.6 Lack of Service Capabilities. Given the reality that China's information technology industry started late, and the industry featured a backward base, China is far behind countries with advanced industries in terms of smart manufacturing upgrading service. The software for upgrade lacks a sound and complete system-wide support. The software products themselves are not mature enough, and even defective, posing high risks in maintenance and upgrade[33, 34]. Meanwhile, the imperfect engineering service market, smart manufacturing upgrade-related service institutions with varied quality and general upgrade solutions result in poor application effects.

3.2. Specific Problems Faced by Upgrade of SMMEs Towards Smart Manufacturing

3.2.1. Equipment Interconnection. In terms of upgrade towards smart manufacturing, SMMEs should give priority to solving the problem of equipment networking and do a good job in developing the Manufacturing Execution System (MES) and Supervisory Control and Data Acquisition (SCADA) based on the internet. With different service lives, the production devices of various SMMEs are often scattered, chaotic and small. Some can be directly connected through intelligent components, while others need to be modified before connecting. In the meantime, there exists such a problem as inconsistent software and hardware interfaces, which not only makes the construction process cumbersome and difficult, but also leads trouble in cost control and great difficulties in upgrade.

3.2.2. Business Digitization. It is another big problem for SMMEs to digitize daily production and management processes and to integrate them into management software according to business characteristics. The businesses of SMMEs vary widely, even though enterprises of the same kind face many differences in management modes. Therefore, the development of management software for small and medium enterprises requires close cooperation between software developers from service providers and business personnel of enterprises. It is also very difficult for the both parties to fully understand each other.

3.2.3. Integration into the Supply Chain. Another core issue for upgrade towards smart manufacturing is to integrate SMMEs into the industrial supply chain in the digital world. SMMEs should connect with upstream and downstream partners based on digitization, and realize data connectivity across the supply chain under the precondition of enterprise information security[35]. Thus, SMMEs will truly become extended workshops of large enterprises, allowing quick and efficient interaction with the latter for synchronous R&D and instant production. Clearly, the dominance of such connection lies in the hands of large enterprises rather than small and medium enterprises in most cases, and the latter will not be accepted by the former if they do not have enough accumulation and improvement in smart manufacturing.

3.2.4. Software Integration. The last problem of smart manufacturing upgrade is how to achieve effective system integration between different software, especially protocols between existing sophisticated software systems and unification of enterprises' database format, which are very critical. The solution requires joint efforts from both the enterprises aiming for upgrade and service providers.

In summary, the main problems facing SMMEs in smart manufacturing upgrade are shown in Chart 1.

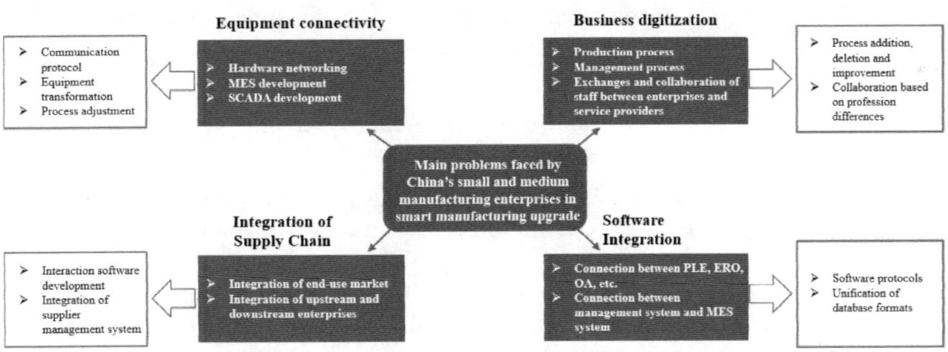

Figure 1. Main problems faced by SMMEs in smart manufacturing upgrade

4. China's SMMEs' measures for smart manufacturing upgrade

Based on the above-mentioned current situation of China's SMMEs and the widespread difficulties in upgrades, as well as the overall situation of China's society and industry, this paper has sorted out and proposed the measures and key issues for China's SMMEs upgrade towards smart manufacturing as follows:

4.1. Overall Division of Jobs of Related Participants

Upgrade of China's SMMEs towards smart manufacturing is a systematic project that requires the participation of all stakeholders and should be implemented as a national strategy. Based on the existing reality that China's SMMEs do not have adequate strength, China should provide specific and efficient support for small and medium enterprises. The government should play a leading role in top-level design and macrostructure to promote the formulation and improvement of relevant policies, laws and regulations. Industrial organizations should fully leverage their own advantages in industrial platforms to assist the implementation of policies and regulations and organize various social resources for smart manufacturing upgrade of small and medium enterprises. Research institutes can conduct in-depth related technical research and project consulting services based on their own research and talent conditions to provide technology and talent support for smart manufacturing upgrade. Finally, as the subject of smart manufacturing upgrade, small and medium enterprises should actively integrate into the wave rather than evade, hesitate or wait. They should do their best to win all available resources according to their own strength, design suitable upgrade solutions and promote upgrade in an active and orderly manner. The division of jobs and cooperation between the relevant subjects in the smart manufacturing upgrade of China's SMMEs are shown in Chart 2.

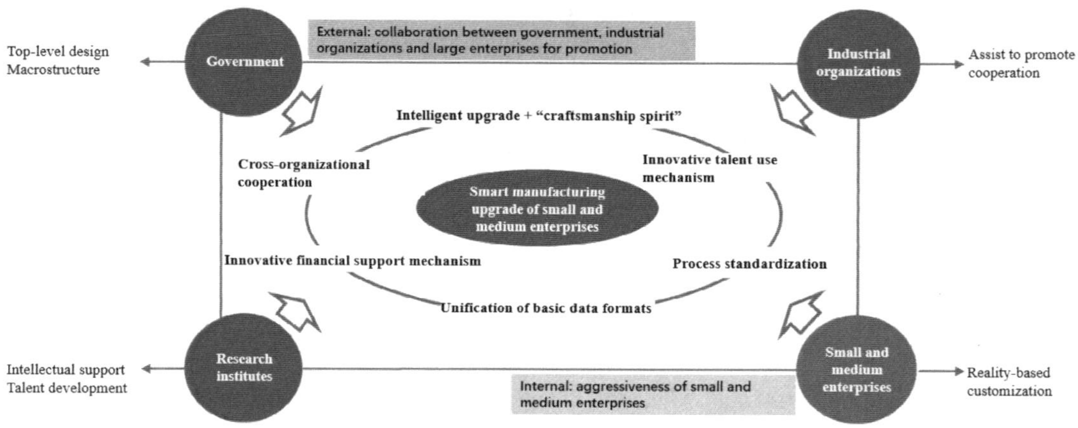

Figure 2. The division of jobs and cooperation between relevant subjects in smart manufacturing upgrade of small and medium

4.2. Supportive Measures Required from the Government and Industrial Organizations

4.2.1. Innovative Financial Mechanism. The expenditure required for smart manufacturing upgrade is a great financial pressure on most SMMEs. Solving such a problem cannot rely solely on enterprises themselves, because the government, industrial organizations, financial institutions and leading enterprises receiving supporting services from many small and medium enterprises are all stakeholders. The government should fully play its role and practically provide as much support as possible for SMMEs in taxation and financial policies. For instance, governments at all levels can provide tax reduction and exemption for enterprises related to smart manufacturing upgrade and coordinate financial resources to provide enterprises with long-term credit loan, financial leasing and so on, which will greatly accelerate the upgrade of enterprises. At the same time, large leading enterprises should also be encouraged to provide certain financial support for supporting small and medium enterprises in shortening the payment days and credit guarantees based on long-term strategic demands and returns.

4.2.2. Promote Cross-Organizational Cooperation. It is significant to carry out cross-organizational cooperation based on requirements of upgrade towards smart manufacturing of SMMEs. It is difficult to change the reality in short term that small and medium e1nterprises are weak in technologies and short of innovation capabilities. Since they have great difficulties in dealing with basic information-based problems, they may have no idea of where to start once encountering new technological, interdisciplinary and cross-field problems. However, these technical problems facing SMMEs during the process of upgrade towards smart manufacturing are often "multiple, small, and scattered" and cannot be skipped, thus requiring more effective support from the external "think tank". Although cooperation between enterprises and research institutions is also promoted at the level of governments and industrial organizations, the mode is generally outdated and inefficient. With high connectivity in today's world, it is possible to give full play to the advantages of the network platform and try new models such as accurate demands, joint R&D and building & operation of virtual innovation space, so as to "melt" practical difficulties in the upgrade towards smart manufacturing of SMMEs. Details are shown in Chart 3.

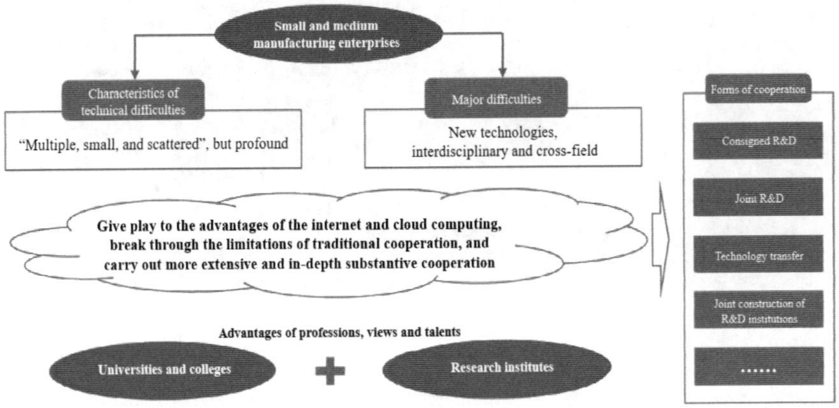

Figure 3. Cross-organizational cooperation during upgrade towards smart manufacturing of SMMEs

4.2.3. Promote the Establishment of a New Talent Use Mode. Although enterprises face varied talent problems, SMMEs have particularly serious problems in such aspect. However, in a highly informationalized era, the government and industry organizations should break the old talent service mechanism and provide "talent sharing" service across the society. SMMEs should seize the opportunity to actively try and fully utilize innovative talent use modes. Certainly, a series of problems, such as which industries and talents are suitable for sharing, how to share knowledge, skills and experience of talents, and how "large, medium and small" enterprises mutually share talents and avoid competition conflicts under the "talent sharing" mode still require more in-depth research and discussion. In order to support the industry transformation and upgrade, governments and industrial organizations at all levels should actively promote top-level design and innovation mechanism, and establish talent use and service modes suitable for smart manufacturing development requirements as soon as possible, which is particularly valuable for the upgrade of SMMEs.

4.3. Upgrade Measures Taken by SMMEs

4.3.1. Promote the Standardization of Main Business Processes. Standardized business process is the foundation for enterprises to promote digitization, informatization and smart manufacturing. Due to limited financial resources and technical level, it is difficult for SMMEs to propose a systematic process normalization suitable for most enterprises in the industry. Moreover, a single enterprise has neither the motivation to promote efficient processes to the outside, nor the authority to formulate standardized process standards. Therefore, in addition to the active promotion of the government and industry organizations, large leading enterprises should make full use of their organizational advantages and management conditions, and effectively unite with software experts to standardize the industry business process. Subsequently, it can be defined in the form of industry recommended standards. Although SMMEs cannot lead this process, they must take an active part and firstly realize the standardization of their own processes.

4.3.2. Achieving the Standardization of Critical Basic Data. After completing internal transformation towards smart manufacturing, manufacturing enterprises must get integrated into the industrial chain as well to realize efficient interaction with upstream and downstream enterprises and the market. Chaotic data format will greatly increase the difficulty of information exchange between enterprises. Similar to process normalization, data standardization should also be guided and promoted by the government and industry organizations. From the perspective of interaction requirements, it is not only necessary to internally achieve data consistency, but also externally. It is necessary for basic data, including enterprise material code, department code, customer code and so on, to be unified in format and to gradually form recommendation standards, as shown in Figure 4. In this process, SMMEs shall not stand by but take an active part and constantly optimize their own data standards.

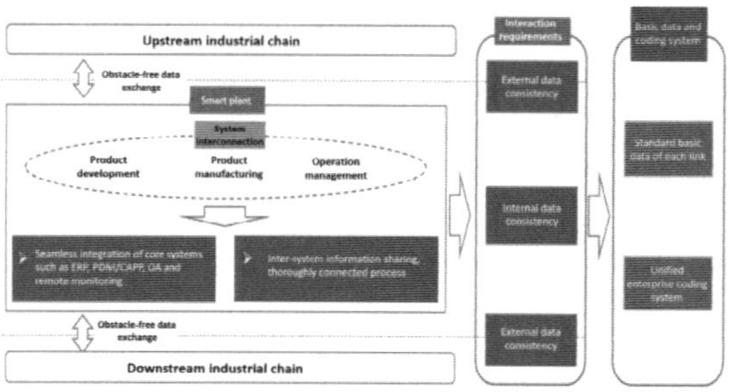

Figure 4. Unification requirements for basic data and enterprise codes

4.3.3. Fostering "Craftsmanship Spirit". Facing upgrade toward smart manufacturing, SMMEs will greatly improve internal operation efficiency and externally achieve good integration with the industrial chain. On this basis, SMMEs still need to stay faithful to their original aspiration, focus on core businesses, commit to specific fields, and continuously optimize and refine technologies and processes to finally provide high-quality personalized products and services and truly become a solid foundation for the smart manufacturing system. In the smart manufacturing system of tomorrow, the importance of "Craftsmanship Spirit" will increase rather than decline. Enterprises ought to be required and guided by the spirit continuously, and make improvement on upgrading and transformation.

4.4. Supportive Measures Required of Scientific Research Institutes

4.4.1. Providing Flexible Scientific Research Cooperation and Intellectual Support. The upgrading of small and medium enterprises towards smart manufacturing is a complex system engineering. In terms of necessary technologies and intellectual resources, SMMEs lack sufficient reserves and are difficult to make large investments. Universities and scientific research institutions have better technical capabilities and rich intellectual resources. They should reach effective cooperation. Previously, only large enterprises drew attention in this kind of industry-college-institute cooperation. Actually, small and medium enterprises have more urgent demands and require more extensive technologies. With increasingly mature information technology conditions, various scientific research institutions can fully cooperate with small and medium enterprise in innovative ways. SMMEs can obtain necessary intellectual support within their financial capacity by improvement and cooperation, as well as sharing of relevant resources through segmentation or integration.

4.4.2. Innovative Talents Co-Development Mode. Talent is the core factor in the process of enterprises' upgrade towards smart manufacturing. Accordingly, the talent training mechanism must be greatly adjusted and innovated to meet the urgent needs of industrial transformation and development in the future. Therefore, universities and scientific research institutes should face the talent demands of numerous SMMEs and try to develop talent co-development modes, and constantly expand the channels for talent interaction with small and medium enterprises, so that the latter that need support most can get continuous output of talents. This is another strategic task for universities and scientific research institutes in this round of scientific and technical revolution.

4.5. Other Key Issues in Upgrade of SMMEs Towards Smart Manufacturing

4.5.1. Timing and Capital Plans for Smart Manufacturing Upgrade. The transformation and upgrade of manufacturing enterprises towards smart manufacturing is a gradual process to integrate information technology into businesses. Enterprises should make good preparations for long-term promotion from the beginning rather than regard it as short-term and periodical work. It is also impossible to "accomplish the whole task at one stroke" by upgrading some hardware and software. At the same time, the implementation process involves almost all business departments inside enterprises, as well as cooperation and communication between enterprises and external service providers. Given that the work is pioneering and tentative, it is very normal to make changes and adjustments in time and capital. For the management of similar upgrade projects, it is inappropriate for enterprises to completely adopt traditional fixed time and capital budget methods but to leave certain flexible space while developing as good systematic and prospective planning as possible.

4.5.2. Conflict of Interest Handling in Smart Manufacturing Upgrade. The process of smart manufacturing upgrade will inevitably lead to significant changes of enterprises' inherent business processes, which will bring a lot of inconvenience to the original business and easily cause resistance of some employees, especially when the process inevitably causes replacement of some manual jobs by automated hardware and software equipment, and fierce resistance of stakeholders. To this end, enterprises must make good preparations in advance for staff placement and persuasion to minimize unnecessary interference during the implementation.

5. Case analysis

We have studied a large number of cases in the upgrade towards smart manufacturing of SMMEs. This paper selects one typical case for brief analysis and explanation, namely the smart manufacturing upgrade of a medium automobile parts enterprise in a southwestern city in China.

5.1. Basic Situation of the Enterprise

As a typical medium enterprise, it has about 800 employees and delivers an annual output of less than 400 million CNY. Its main products include automotive sheet metal stamping and welding parts, provided for Changan Ford and Changan Automobile. In order to break through the bottlenecks hindering development and become a first-class auto parts supplier, the enterprise has implemented a smart manufacturing upgrade project since 2015.

5.2. Main Content of Upgrade

During the upgrade, the enterprise built a dedicated enterprise local area network (LAN) to isolate from the external public network with gateway software. The network is preset physical isolation measures in case of emergency, and connect PLC, industrial robots, CNC and other automation equipment to the public data collection platform for normal collection of data. A dedicated database was established and a post-processing data platform was developed. For the risk of quality defects caused by mold damage in some key process links, the enterprise deploys an optical automatic detection system and connect it to the public data platform. In addition, the key equipment control system was upgraded for safe control and product quality traceability.

5.3. Implementation Effects of Upgrade

The intelligent system delivered remarkable results soon after it was put into operation. The state of isolation between equipment in the past was completely broken, which allows the management layer to monitor the running status of all equipment in real time, to substantially reduce the waste of materials and to more flexibly switch between products. Some important products even allow automatic quality monitoring, with much smaller difficulties in quality maintenance and problem traceability.

5.4. Support of the Government and Industry Organizations During the Upgrade

In order to reduce the burden of upgrade, the enterprise successfully applied to the local government for the subsidy for intelligent upgrade project during the process of implementation, which ensures the success of the upgrade. For some tough problems, the enterprise actively participated in many special lectures and consultations organized by the local industry alliance organizations with experts from the upstream and downstream of the industry.

5.5. The Enterprise's Own Efforts During Upgrade

During the upgrade towards smart manufacturing, the enterprise leadership paid great attention to it and specially set up an information business unit. In addition, technology, production, quality, logistics, finance and other relevant departments are required to provide support during the whole process, and the responsible department should regularly report to the top leaders of the enterprise. In order to deploy intelligent production and management systems, the front-line employees of relevant service providers work together with the employees of all departments of the enterprise under the coordination of the information business unit to comprehensively sort out the process and initially standardized the process. In view of the chaotic data formats and codes of different departments, the enterprise also standardized data for unification. In order to minimize the impacts of the deployment of intelligent systems on the established production plans, the construction personnel and production personnel cooperated well during the installation process of hardware and software. In addition, for the changes of positions and responsibilities caused by the operation of the new system, the enterprise has fully explained and guided relevant employees in advance and made proper resettlement afterwards.

5.6. Cooperation with Research Institutes During Upgrade

In the process of system integration and personnel training, the enterprise actively carried out in-depth cooperation with relevant experts from local universities and a famous engineering university in Beijing, which greatly reduced the burden on system function integration and personnel training.

In summary, the smart manufacturing upgrade of the enterprise is very typical. The local government, industry organizations, enterprises and some universities have participated in this process, with reasonable division of jobs and close cooperation. Great importance has been attached to the implementation of the project from front-line employees to the top leadership. Although the project has gone through all kinds of unexpected problems and setbacks, it has finally achieved the expected goal. Certainly, as a traditional small and medium manufacturing enterprise initially attempting smart manufacturing upgrade, its work is not that perfect, and there is still a gap between the performance and the ideal state of upgraded system functions. However, it has provided a basic prototype of smart plant for the enterprise and laid foundation for the gradual and continuous improvement in the future. This case shows that as long as SMMEs in China have explicit objectives, clear ideas, firm implementation, and down-to-earth practice, it is completely possible to make great breakthrough in smart manufacturing upgrade under existing conditions.

6. Conclusion

SMMEs are related to the overall development of the manufacturing industry, especially in the context of the global manufacturing industry upgrading towards smart manufacturing, the status of China's SMMEs will become more prominent, and the demands for smart manufacturing will become more urgent. In the future, excellent SMMEs should achieve intelligent production internally, get fully integrated into the industrial chain externally, and build a bridge between the market and upstream and downstream digitally to achieve efficient interaction with other links in the industrial chain. At present, China's SMMEs face many challenges in the upgrade towards smart manufacturing, such as common problems in business processes, software and hardware, talents and capitals, as well as specific difficulties in equipment interconnection, process standardization, data standardization, software system integration and so on. Therefore, this paper advises: the government, industry organizations, small and medium enterprises and scientific research institutes should have effective division of jobs and mutual cooperation in the process of upgrading towards smart manufacturing of SMMEs. Governments and industry organizations should carry out top-level planning and major promotion in developing new financial systems, promoting trans-institutional cooperation and developing new talents use modes. Colleges and universities and scientific research institutes should give full play to their advantages in technology and intellectual resources, break through the limitations of traditional cooperation, and explore innovative collaboration modes that provide direct support to SMMEs. SMMEs should actively take the initiative to meet this tide of upgrading, focus on promoting the standardization of process and data format, and integrate the cultivation of "craftsmanship spirit" into the upgrade towards smart manufacturing. Finally, SMMEs should know that the upgrade towards smart manufacturing is time consuming, systematic and uncertain, requiring thorough consideration of the possibility of time and capital changes in advance, as well as resettlement and treatment of employees who may suffer from such changes.

Reference

[1] Esmaeilian B, Behdad S and Wang B 2016 *J. M. Sys.* **39** 79
[2] Zhou K, Liu T and Liang L 2016 Int. J. Manuf. Res. 11 167
[3] Lu H and Weng C 2018 Tech. Forec. Soc. Chang. 133 85
[4] Liu W 2010 J. DONGBEI Univ. Fin. and Econ. 6 52
[5] Hu R and Zhao H 2005 Chin. Manage. Inf. 12 47
[6] Li L and Zeng M 2011 Chin. Col. Eco. 16 23
[7] Zhao F, Liu Z and Shi T 2017 Sci. Tec. Prog. Pol, 14 85
[8] Zhao F and Liu Z 2016 For. Sci. Tec. Chin. 1 58
[9] Shao A 2017 Eco. Rev, 1,74
[10] Xiao D and Hou S 2017 Chin. J. Manage. Sci. 4 86
[11] Liu Z, Chen J and Hao H 2017 Eco. Rev. 9 105
[12] Lan F and Liu X 2017 Int. J. Tec. Pol Manage. 17 360
[13] Sun J 2012 Chin. Bus. (Theor. Res.) 11 24
[14] Fan R 2013 Jiangsu Commer. Forum. 3 72
[15] Leitao P, Mendes J and Bepperling A 2012 J. Int. Manuf. 23 2551
[16] Lee J, Bagheri B and Hung-An K 2015 Manufacturing Letters. 3 18-23.
[17] Ruiz N, Giret A and Botti V 2014 Comput. Ind. Eng. 76 148
[18] Zhou F, Lin X and Luo X 2018 J. Vis. Lang. Comput. 44 58
[19] Yao A and Lin R 2015 Appl. Mech. Mat. 789 1082
[20] Fisher O, Watson N and Porcu L 2018 J. Manuf. Sys. 47 53
[21] Liu X, Shahriar M and Sunny S 2017 Journal of Manufacturing Systems. 43 352-364.
[22] Giordani S, Lujak M and F. Martinelli 2013 Comput. Indust. Eng. 64 19
[23] Du C and Du X 2017 Eco. Rev. 4 110
[24] Zhao F, Su R and Liu Z 2018 Supply Chain And Auto Power (Chin. Mach. Press)
[25] Liu Z 2017 Zhao Fuquan's Insights on the Automotive Industry (Chin. Mach. Press)
[26] Moeuf A, Tamayo S and Lamouri S 2016 IFAC-PapersOnLine. 49 71
[27] Liu F 2013 Guide Bus. 1 212

[28] Wang H 2012 Prod. Res. 4 228
[29] Zhao L and Yu Q 2008 Manag. Obs. 18 59
[30] Zhao F, Su R and Liu Z 2016 Insights for a Stronger Auto Industry (Chin. Mach. Press)
[31] Wu X 2018 Ent. Ref. Manage. 1 116
[32] Yi W 2012 Ent. Vit. 3 88
[33] Liu F 2010 Ent. Eco. 1 42
[34] Tuptuk N and Hailes S 2018 J. Manuf. Sys. 47 93
[35] Helo P, Suorsa M and Hao Y 2014 Comput. Indus. 4 646

IOP Conf. Series: Materials Science and Engineering **627** (2019) 012012 doi:10.1088/1757-899X/627/1/012012

Economic prospects of multiproduct manufacturing based on the Afrikanda deposit ores

Vladimir Masloboev, Sergey Fedoseev, Pavel Tcvetkov, Sergey Krivovichev

Federal Research Centre "Kola Science Centre of the Russian Academy of Sciences", 24a Fersman street, Apatity, Russia

pscvetkov@yandex.ru

Abstract. This article describes the results of the economic efficiency assessment of the Afrikanda deposit (Russia, Murmansk Region) development based on the Monte Carlo method. The project was divided into two phases: Phase 1 – production of perovskite concentrate, phase 2 - production of TiO_2, Nb_2O_5, Ta_2O_5 and REE concentrate. Comparative evaluation was conducted for two options: only phase 1 and phase 1 + phase 2. Obtained results showed that the profitability as well as the payback period of the first option, which is less capital-intensive, can be in 1.5-2 higher. Despite this, the NPV of the first option is 5 times lower than the second, and there are prospects for further growing of this gap beyond the 15-year evaluation period. It was concluded that there are significant economic opportunities of the integrated deposit development, which may be implemented with additional measures of state support.

1. Introduction

Currently, the world markets are experiencing a shortage of rare earth elements (REE), which hinders the development of several industries, including chemical engineering, shipbuilding, aerospace, nuclear power, production of catalysts, ceramics, paints, fiber glass, superconducting materials etc [1]. This is also true for Russia, which produces almost no REE, despite its significant mineral resources. The involvement of non-traditional raw materials in the processing can serve as a solution to this problem [2]. One of such raw materials is perovskite, which can be extracted from the ores of the Afrikanda deposit [3].

The deposit is located in Murmansk region 1.5 km from the railway station (location: 67°26'12.8"N 32°44'54.8"E). It has an area of 4.5 km^2, explored only on 420 m in depth. The lower boundary of the field is not reached. The field is connected by rail with the industrial centers of the country, Murmansk and Kandalaksha sea ports. The area is provided with electricity (Kola nuclear power plant) and water resources (lake Imandra). Near the deposit is a residential village. Total ore reserves in the field are estimated at 626.2 million tons, approximately. In addition to the presence of REE, the main useful components contained in these ores are titanium, niobium and tantalum. The total amount of TiO_2 in the field is estimated at 52.2 million tons (average content - 9.2% [3]).

Production of titanium dioxide in Russia began only in 2014 and today is insignificant, despite the fact that the country's raw materials reserves are among the largest in the world, and the volume of the domestic market is almost unlimited. In this regard, Russian industries use imported raw materials, obtained mainly from China, as well as Ukraine and the United States. Taking into account political situation, it is extremely negative dependence [4].

The Afrikanda deposit has a relatively long history [5]. In Soviet period the question was raised about the development of the field for the needs of ferrous metallurgy, and in 1957 a processing plant and a quarry were created. 11 thousand tons of perovskite were extracted and used for the production of ferrotitanium. Some parts of this volume were transferred to research institutes of the USSR and in other countries. Due to the low technological efficiency of dressing process and unpreparedness of enterprises for the production and processing of ferroalloys, the production of perovskite concentrate was stopped and in 1982 the project was closed.

To date, an effective flotation scheme for the processing of perovskite-titanium magnetite ores with a closed water circulation has been developed at the Kola Scientific Center of the Russian Academy of Science (KSC RAS) [6]. On the basis of this technology a conceptual three-stage model of the Afrikanda deposit development was proposed (Figure 1). The third stage is formulated only as a concept; as further research is required to improve the efficiency of REE concentrate processing.

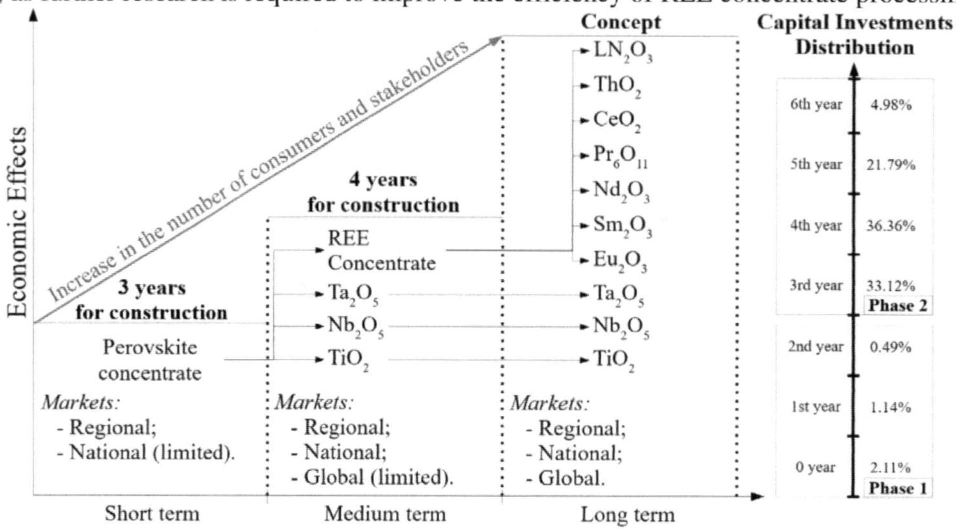

Figure 1. Conceptual model of the Afrikanda deposit development.

Figure 2, in addition to the conceptual model of production, shows the distribution of capital investments by years for the first two stages of the enterprise development, according to the calculations of the service mining company LLC "ArcMineral" [7]. The shown distribution of investments assumes that there will be no time gap between the implementation of the first and second phases, i.e. these phases are considered as a single project.

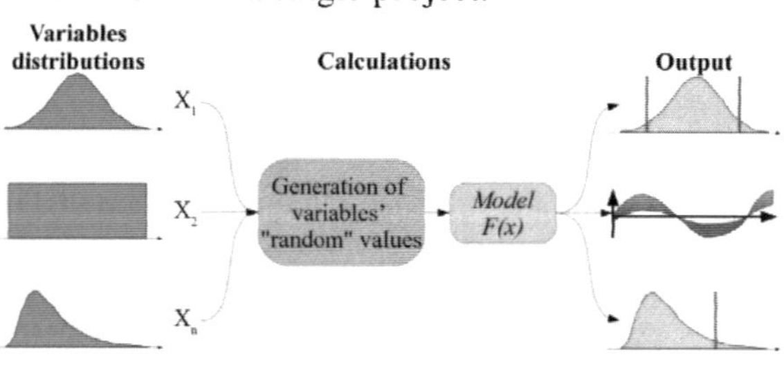

Figure 2. Basics of Monte-Carlo method.

The purpose of this article is to assess economic efficiency of the first two stages of the proposed model, which are based on the use of nitric acid technology for perovskite concentrate production and processing. This assessment will also show the feasibility of the second Phase of the enterprise development.

2. Materials and Methods

2.1. Assessment method

The assessment was based on the Monte Carlo simulation method. This method is widely used for evaluation of mineral deposits, as it allows taking into account the influence of many uncertain factors on the development process, including geological, technological and market.

The essence of the method (Figure 2) consists in statistical modeling of random functions and random variables to determine the characteristics of their distributions. The idea of the method is that instead of analytical describing of a process or phenomenon, its "drawing" is carried out with a special procedure that gives a random result in a certain range. As a result of multiple implementations of such drawings, the resulting data set can be used as an artificial statistical material that can be processed by standard mathematical methods.

2.2. *Initial data and design of the study*

Information from various studies conducted by the KSC RAS on the possibility of Afrikanda deposit ores processing were taken as initial technical data (Figure 3). The assessment also takes into account available infrastructure of the Soviet period, which can be effectively used [8]. Capital investment were taken from the assessment of LLC "ArcMineral" [7] and were increased by the inflation rates in Russian Federation in 2016-2018. It is also taken into account the presence of a power plant that can provide energy to the enterprise [Masloboev, personal communication[1]].

Basic (probable) prices for products were taken according to [7]. Taking into account high volatility of the products prices, top and bottom thresholds were taken as minimum and maximum prices in Russian markets in 2018 [Skripachev, personal communication[2]] (Table 1).

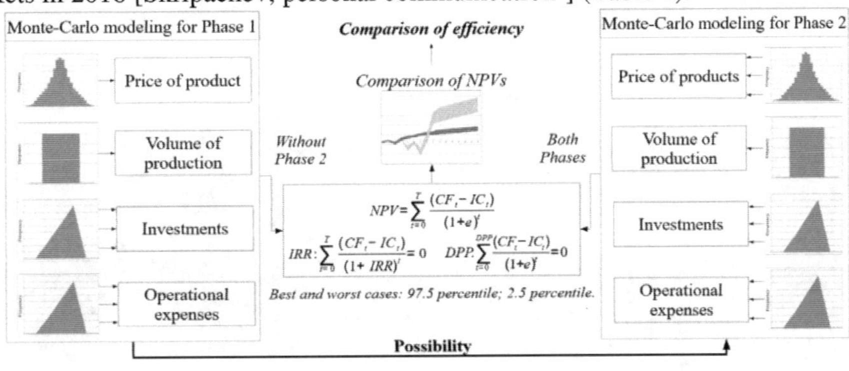

Figure 3. Design of the study.

Table 1. Price ranges for products.

Product	Units	Probable price	Min	Max
Perovskite concentrate	USD/ ton	300	282.8	321
TiO_2		1500	1425.4	1595.2
REE concentrate		6000	5784.1	6331.1
Nb2O5		50000	49491.7	51030
Ta2O5		150000	148895.6	151770.3

[1] Masloboev, V.A. Kola chemistry and technological cluster. Kola Federal Research Centre, Russian Academy of Science (2018) (*In Rus.*)

[2] Skripachev, S.Yu. On the state and prospects of Rare Earth Elements. Presentation at the meeting on the prospects for the development of the rare and rare earth metals industry under the leadership of the Deputy Minister of industry and trade of the Russian Federation, All-Russian Federal Institute of Aviation Materials (November 2018) (*In Rus.*)

According to [7], production volume of the plant in terms of perovskite concentrate will be 170 000 tons per year with possible volatility of 5%, depending on geological features. With such production capacity, the reserves of the field [3] will last for more than 200 years. This fact shows the specific feature of such projects – the evaluation period significantly depends on the available reserves of the deposit and, as a rule, lies between 10 and 20 years [9, 10, 11]. Thus, in this study the evaluation period is assumed to be 15 years.
Risk-adjusted discount rate (15%) was taken as a sum of inflation rate (4.3 %), Bank of Russia key rate (7.5 %) and risk rate (3.2 % [12]). Depreciation method - linear, term - 15 years (equal to evaluation period). Design of the study is shown in Figure 3, where: NPV - Net Present Value, IRR – Internal Rate of Return, DPP – Discounted Payback Period, CF – Cash Flows, IC – Capital Investments, e – discount rate, t – period.

3. Results

Based on the Monte Carlo simulations, the possible values of NPV for Afrikanda project were calculated. The evaluation was carried out for two options of the project development. The first option involves only the production of perovskite concentrate (Phase 1 – Figure 4-A). The second option involves, in addition to phase 1, the construction of a dressing plant for the production of TiO_2, Nb_2O_5, Ta_2O_5 and REE concentrate from perovskite concentrate (Phase 2 – Figure 4-B).
From the obtained distributions, the best (97.5 percentile) and the worst (2.5 percentile) NPV variants were determined. On the basis of these data, the comparative analysis of NPVs was carried out (Figure 5).

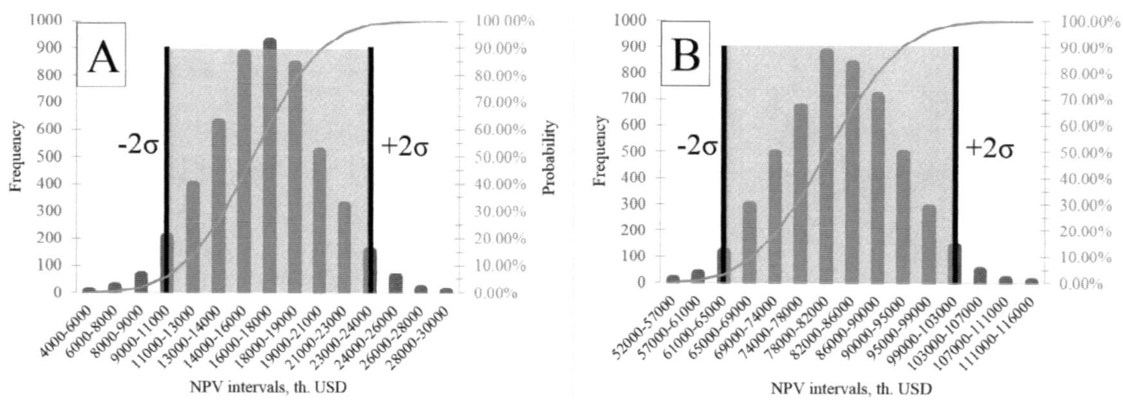

Figure 4. Probability and frequency of NPV distribution for Option 1 (A) and Option 2 (B).

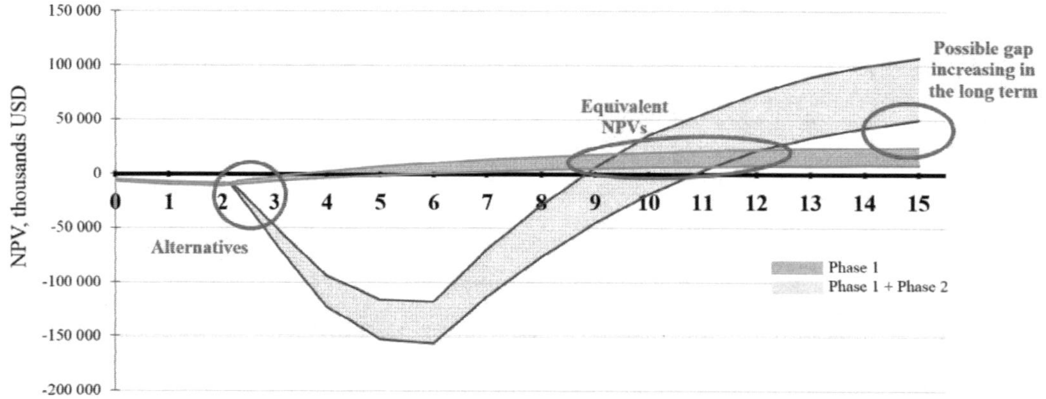

Figure 5. Comparison of Phase 1 and both Phases NPVs.

The main idea behind Figure 5 is to show that more capital-intensive project could provide much higher economic effect in the long term. As it was proposed in [7], there is no time gap between two phases. Phase 2 is beginning at the end of the third year and drops NPV below 150 mln USD in the 6th and 7th years of the project execution, when single Phase 1 could already pass payback period. However, when Phase 2 pass its payback period in 10th - 12th years, its NPV rises significantly faster than have Phase 1. It allows to make a highly probable prediction about further increasing in gap between these two options' NPVs in the future. Analysis of NPVs content in the best and worst cases allowed to determine the ranges of key performance indicators of the investment project (Table 2).

Table 2. Price ranges for products.

Indicators	Units	Phase 1		Phase 1 + Phase 2	
		Worst	Best	Worst	Best
Net Present Value	th. USD	8 284	25 192	59 922	127 182
Discounted Payback Period	years	6.50	4.44	11.86	9.80
Internal Rate of Return	fract. of units	31.3%	46.6%	21.6%	25.7%
Index of Profitability	fract. of units	2.01	3.49	1.37	1.65

4. Discussion

The following results were obtained in this study:

1. Both considered options for the deposit development are cost-effective, as evidenced by the distribution of NPVs (Figure 4). Thus, the technology of KSC RAS for perovskite processing has a significant economic potential and can be a key factor that will ensure the sustainable operation of the enterprise in the long term. This is also due to the high demand for final products both in the world and in the domestic market of Russia.
2. Despite the fact that potential economic effect (Figure 4) from the implementation of Phase 2 is significantly higher than simple production of perovskite concentrate, the performance of the project within the 15-year period decreased (PI, IRR – Table 2). In addition, the payback period of the project doubled, which can become a barrier to attract investments for the second phase. However, the higher efficiency of the first phase may be an illusion, as a shortage of titanium-containing raw materials is observed in the domestic market of Russia, but not in the surrounding regions. Dependence on one market and one final product can have a negative impact on the financial stability of the enterprise.
3. The negative perception of the long payback period of the project can be mitigated by postponing the start of Phase 2 for some years. It will make possible to completely separate phases into two independent projects (from financial point of view), as well as to update information about market conditions and necessary investment exactly before implementation. In addition, the implementation of Phase 2 after passing the payback border of phase 1, will allow to use the available net profit of the enterprise and to reduce the amount of borrowed funds, if necessary.

5. Conclusions

This paper presents the analysis of economic efficiency of the Afrikanda deposit development, based on the Monte-Carlo method. The choice of the method is associated with the lack of clear initial data and a lack of knowledge of both the deposit geological features and the potential ore processing technology.

The results allow to make an optimistic forecast regarding the economic potential of the field. However, given the difficult economic situation in Russia, the implementation of such capital-intensive projects is complicated. It seems appropriate to use the mechanisms of public-private partnership for the project implementation. Moreover, it is necessary to develop measures of state regulation of subsoil use, which will make capital-intensive projects of integrated field development more attractive to investors and other stakeholders.

In this regard, further research will be focused, firstly, on the development of economic measures to increase the investment attractiveness of large-scale deposits development in the Russian Arctic.

Secondly, on the development of stakeholders interaction mechanism in the framework of Afrikanda project, including the definition of their key areas of responsibility and main interests.

6. Acknowledgments

The project was carried out within the framework of state order No. 0226-2019-0024 "Development of scientific bases and substantiation of eco-economically balanced resource-saving integrated development of natural resources in the Arctic zone of Russia", IEP KSC RAS.

7. References

[1] Didenko, N. and Kunze, K. 2015. Modeling the economic growth of Arctic regions in Russia. *Computational Methods and Experimental Measurements* XVII, 59 (2015), p. 179.

[2] Kozlov, A.V., Gutman, S.S., Zaychenko, I.M. and Rytova, E.V. (2015) The strategy for the development of the Arctic zone of the Russian Federation through the concept of a regional indicators system. *Mediterranean Journal of Social Sciences*, 6 (5 S4), p. 379.

[3] Kalashnikov, A.O., Konopleva, N.G., Pakhomovsky, Y.A. and Ivanyuk, G.Y. 2016. Rare earth deposits of the Murmansk region, Russia—a review. *Economic Geology*, 111(7), 1529-1559.

[4] Vyboldina, E., Cherepovitsyn, A., Fedoseev, S. and Tsvetkov, P. 2016. Analysis of export restrictions and their impact on metals world markets. *Indian journal of science and technology*, 9(5), 87633.

[5] Yudin, B.A. and Zak, S.I. 1971. Titanium deposits of northwestern USSR (eastern part of Baltic Shield). *International Geology Review*, 13(6), 864-872.

[6] Mudruk, N.V., Korovina, Yu.V., Nikolaev, A.I. and Safonova, L.A. 2018. Extraction of rare-earth elements from the solutions after perovskite concentrate decomposition by nitric acid. *Proceedings of the Kola science center of RAS*, 323-327. DOI= 10.25702/KSC.2307-5252.2018.9.1.323-327

[7] Trenin, A.D. 2016. Afrikanda Project - Investors Presentation. Prepared by ArcMineral LLC, Murmansk Region (2016). URL= http://www.osatom.ru/mediafiles/u/files/News/ Atomexpo%202016/2016.05.30_01_Afrikanda_Trenin.pdf (Accessed: 17.06.2019) (*In Rus.*)

[8] Kosarev, O.V., Tcvetkov, P.S., Makhovikov, A.B., Vodkaylo, E.G., Zulin, V.A. and Bykasov, D.A. 2018. Modeling of Industrial IoT complex for underground space scanning on the base of Arduino platform. In *Topical Issues of Rational Use of Natural Resources: Proceedings of the International Forum-Contest of Young Researchers* (April 18-20, St. Petersburg, Russia, 2018), p. 407. CRC Press.

[9] Akkoyun, O. 2012. Simulation-based investment appraisal and risk analysis of natural building stone deposits. *Construction and Building Materials*, 31, 326–333. doi= 10.1016/j.conbuildmat.2012.01.003

[10] Krohns, S., Lunkenheimer, P., Meissner, S., Reller, A., Gleich, B., Rathgeber, A., ... Loidl, A. 2011. The route to resource-efficient novel materials. *Nature Materials*, 10(12), 899–901. doi= 10.1038/nmat3180

[11] Asad, M.W.A. and Topal, E. 2011. Net present value maximization model for optimum cut-off grade policy of open pit mining operations. *Journal of the Southern African Institute of Mining and Metallurgy*, 111(11), 741-750.

[12] Ponomarenko, T. V., Cherepovitsyn, A. E., Fedoseev, S. V. and Sidorov, D. E. 2016. Organizational-economic mechanism of financing strategic investment projects at the regional level in regions with poor infrastructure. *International Journal of Applied Engineering Research*, 11(16), 9007-9013.

The Automatic Measurement System for Harmonic-drive Devices

Yuansong Zheng[1], Zhifeng Lou[1, 2], Xiaodong Wang[1, 2, a] and Siying Ling[1, 2]

[1] Key Laboratory for Micro/Nano Technology and System of Liaoning Province, Dalian University of Technology, Dalian 116024, China

[2] Key Laboratory for Precision and Non-traditional Machining of Ministry of Education, Dalian University of Technology, Dalian 116024, China

Email: xdwang@dlut.edu.cn

Abstract. Harmonic-drive transmission finds many applications in industry, especially in manufacturing of industrial robots. As the key devices of an industrial robot, the harmonic drive reducers determine the robot's performance, which includes the positioning accuracy, loading capacity, life time, etc. To satisfy the need of manufacturing high performance harmonic drive devices, development of competent measurement system is of great importance. An automatic measurement system with high-precision and multifunction is under development. The system consists of the power module, loading module, torque and rotation measuring module. The measurement and operation principle were explained. The factors affecting the accuracy of the measurement system were analysed. The control system was designed to realize the automatic loading and acquisition of torque and rotation data. The principle and measurement software of the control system were also introduced. Finally, a harmonic reducer with the gear ratio 120 was measured to verify the feasibility of the measurement system.

1. Introduction

Harmonic drive, occasionally called "strain-wave gearing", employs a continuous deflection wave along flexspline to transfer motion and power. Because of its unique transmission principle, it has some advantages such as large gear ratio, high transmission precision, high torque capacity and efficiency, lightweight and compact dimensions. Due to these excellent performances, harmonic drive has been widely applied in industrial robots, measuring instrument, assembly equipment, national defense industry, aerospace, and aircraft system.

With the wide application of harmonic drive, the transmission error and efficiency have enjoyed widespread attention from researchers. According to the conventional gear drive model, C.A.Shuwalov predicted the transmission error of harmonic drive by considering the error of wave-generator [1]. By translating manufacturing and assembly error into overall transmission error, Emel'yanov developed a detailed model to calculate position error [2]. Based on the operating principle of harmonic drive, Xuefeng X drawn such a conclusion that transmission error is directly affected by the error of gear pair, and the error of wave-generator cause transmission error through affecting the deflection of flexspline, an model was establish to calculate the transmission error [3]. Depended on geometric models of gear-tooth rubbing, Nye developed static and kinematic models to

estimate mechanical efficiency [4]. By calculating the driving power and friction power, Jianchu Y proposed a model to obtain the meshing efficiency of involute cylindrical gear [5].
Affected by many complex factors, the theoretical results calculated by these models are always inaccurate, thus, using an experimental method to measure the transmission performance is necessary. Based on rotary encoder and 120-tooth high-precision gear, Yanabe S developed a measurement system to obtain transmission error, however, this instrument can only be used for static measurement [6]. According to the electrical and optical angle measurement method, Weiyong Li designed a high precision and automatic angle-measuring platform to measure transmission error and backlash [7]. Hejny S W designed a test platform for the study of harmonic drive, observation were also made concerning the way in which flexibility effects the magnitude of the transmission error, in addition, the proportional-derivative method were implemented in both motor state and load state feedback control, and the load state feedback effort was successfully compensated for the effects of transmission error during the regulation of the load position [8]. Aiming at the transmission efficiency, Jianhong R developed a measurement system, both torque motor and magnetic powder brake were used to achieve loading with wide range and high precision [9]. Based on virtual instrument technology, Zhen C designed a mechanical efficiency measuring system, and modular design concept was used to improve the automatic level [10].
The above instrument can only be used to measure one kind of transmission performance; therefore, the development of multifunction integrated test system with general-purpose and high automaticity has received widespread attention. Qi L designed a measurement system which can be used to measure transmission accuracy, stiffness, backlash, mechanical efficiency of the RV reducer with different series and specifications [11]. Based on Permanent Magnet Synchronous Motor system, optical-electricity encoder, magnetic powder brake and corresponding resistant torque master, Hongchun W design a dynamic and static test system to measure transmission error, carrying ability and output rotational speed of harmonic drive [12]. Based on this research, Xuesong M developed a measuring system which can be used to capture mechanical efficiency, transmission error, service life, overload capacity, starting torque and stiffness [13]. Although these instruments can be used to measure a variety of performance, the accuracy is not enough.
In this paper, according to the measuring principle of transmission error and mechanical efficiency, an automatic measurement system for harmonic drive transmission performance was developed, and the principal factors which may affect the accuracy of measuring platform were analysed. In addition, the control system based on modular design was introduced. Finally, measurement experiments were carried out using this system to verify the feasibility of the measurement system.

2. The measurement system

Transmission error and mechanical efficiency are important indicators of harmonic drive's performance, test parameters for calculating these two features include torque, rotation at input and output of harmonic drive. According to these parameters, transmission error and mechanical efficiency can be obtained through the method described as follows.

2.1. The transmission error

Transmission error θ, is typically measured by subtracting the rotation at the output of the harmonic drive from the ideal rotation which is obtained through the input rotation divide by the ideal gear ratio, the expression can be described as follows:

$$\theta = \frac{\theta_i}{i} - \theta_o \tag{1}$$

Where i is the ideal gear ratio, θ_i and θ_o are rotation at the input and output of the harmonic drive.
Both static transmission error and dynamic transmission error are used to evaluate transmission accuracy. When the input shaft rotates a certain angle, the rotation at the output of harmonic drive can be measured under stationary state. According to equation (1), static transmission error could be calculated. Static transmission error does not consider the influence of inertial load, so it cannot reflect the dynamic characteristics. The dynamic transmission error is calculated by measuring the input and

output rotation of harmonic drive during its operation progress, it is more consistent with the actual working condition, therefore, this method was adapted in this research. A high precision rotary encoder was used to obtain the rotation at the input and output of harmonic drive, and transmission error was calculated based on equation (1).

2.2. The mechanical efficiency

Due to gear-tooth-friction and damping of wave-generator, harmonic drive exhibit a finite power loss, therefore, it is of great significance to study the mechanical efficiency of harmonic drive.
Mechanical efficiency η , is calculated by the power at the input and output of the harmonic drive, the expression can be described as follows:

$$\eta = \frac{P_o}{P_i} = \frac{T_o}{iT_i} \tag{2}$$

In equation (2), P_o and P_i are the power at input and output of harmonic drive, i is the ideal gear ratio, T_o and T_i are torque at input and output of harmonic drive. In order to calculate mechanical efficiency, two high-precision torque sensors were used to obtain torque.
According to the above analysed, rotary encoder and torque sensor are the key components to measure transmission error and mechanical efficiency. According to this principle, a measurement system was developed.

2.3. The structure of the measurement system

The structure of the measurement system was shown in figure 1. It was composed of motor, rotary encoder, torque sensor, magnetic powder brake, and pedestal. Torque sensor and rotary encoder were used to acquire torque and rotation at the input and out of harmonic reducer, the motor provides power for this system, in order to conform to the actual working condition, magnetic powder brake was used to impose different load during measurement progress.

2.4. Analysis of the influence of angular deviation

As shown in figure 1, harmonic reducer and rotary encoder were connected through diaphragm coupling, therefore, the existence of coaxiality error is inevitable, it is necessary to analyse the relationship between coaxiality error and transmission error of diaphragm coupling.
Based on simulation method, magnitude of diaphragm coupling transmission error were present in this research when the angular deviation of the connecting shaft is 0 degree, 0.1degree, 0.2degree and 0.3 degree. The simulation results were presented in figure 2. As shown in figure 2, when angular deviation is 0.3 degree, the transmission error is ±7", it would have a great affection on measurement precision. The transmission error is ±0.8" when angular deviation is 0.1 degree, in this case, the transmission error caused by misalignment of diaphragm coupling could be ignored. Therefore, the angular deviation of diaphragm coupling between the rotary encoder and harmonic reducer should be within 0.1 degree.

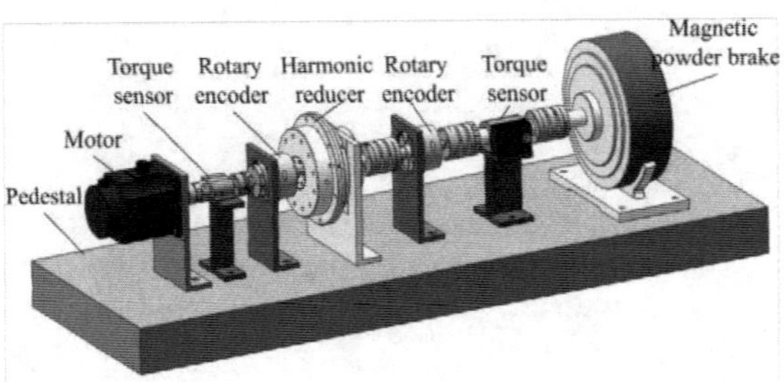

Figure 1. Structure of the measurement system

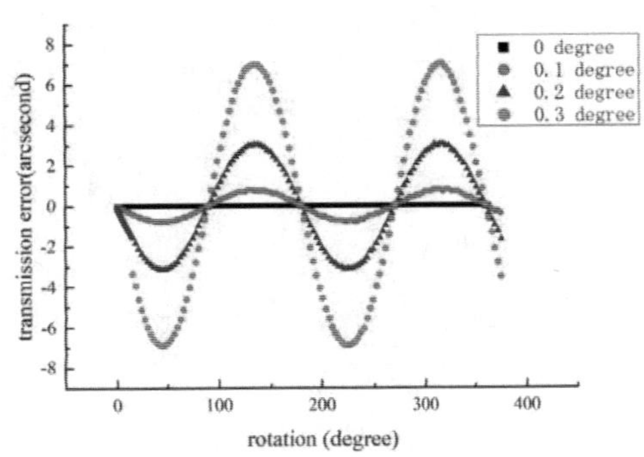

Figure 2. Transmission error of diaphragm coupling in different angular deviation

As shown in figure 1, all shafts were contacted through couplings, if the shafts were connected from left to right during the building progress, coaxiality error of measurement system would be accumulated. In order to solve this problem, first, the motor was fixed on the pedestal, and then components were installed from right to left. Because the motor shaft was always used as a reference during the progress of building the platform, the cumulative error could be minimized as much as possible.

According to the above analysis, the measurement system was built up (see figure 3). A motor (HG-SR102J, made by Mitsubishi, rated speed 2000RPM) and servo driver consist of the motor driver module. Torque sensors (0261E, with accuracy 0.1%F.S, repeatability ±0.02% F.S, Nominal Torque 5N.m; 0261Due, with accuracy 0.1%F.S, repeatability ±0.02% F.S, Nominal Torque 20/200N.m, made by Lorenz Messtechnik GmbH) were used to capture the torque at the input and output of the harmonic reducer. The rotation at input and output of harmonic reducer were obtained by the high-precision rotary encoders (RCN2310, system accuracy ±5 arc-second, 26-Bit, RCN2510, system accuracy ±2.5 arc-second, 28-Bit, made by Heidenhain). Magnetic powder brake (CZ-20, Nantong Hailing Magnetic Brake Manufacturing CO.LTD) provide a load for the measurement system.

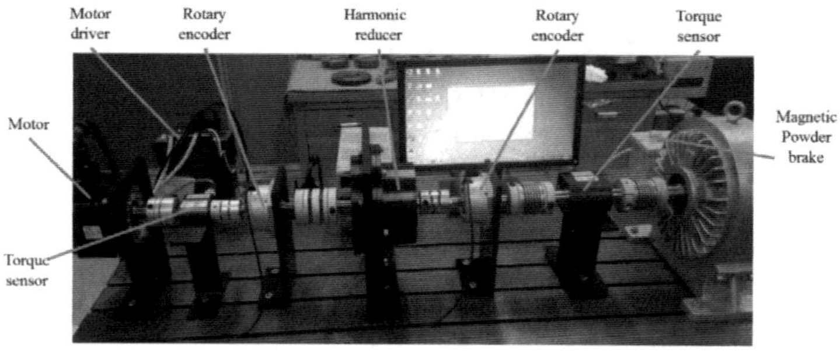

Figure 3. The measurement system

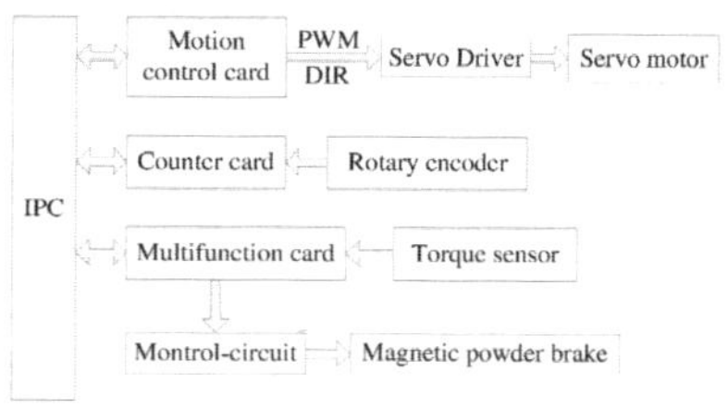

Figure 4. Principle of control system

3. Automatic control of measurement

The control system was composed of servo motor drive module, data acquisition module, loaded module, IPC and software, principle of the control system was shown in figure 4.

According to the parameters of direction and speed, DIR and PWM signals were output from the IPC to adjust the rotation of motor, and transmission performances of harmonic reducer were tested under different speed. Data acquisition module was composed of PCI-1710 data acquisition card (12-Bit, sampling rates 100 KHz, made by Advantech), torque sensor, rotary encoder, and IK220 counter card (made by Heidenhain). When the reducer was driven by motor, PCI-1710 was used to acquire the torque, and rotation was obtained by IK220 card. PCI-1710 data acquisition card, conditioning circuit, and magnetic powder brake consist of the loading module, PCI-1710 send analog quantity to conditioning circuit to control magnitude of the load.

Measurement software, consists of three threads, is mainly used for motor control, data acquisition and processing, and loading control. Flower chart of the measurement software was shown in figure 5.

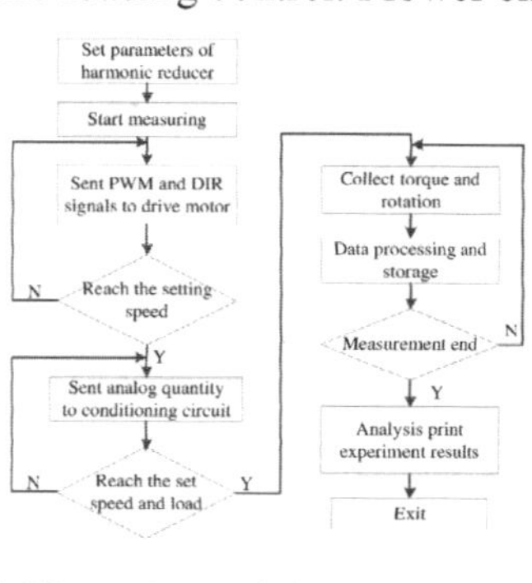

Figure 5. Flow chart of the measurement system

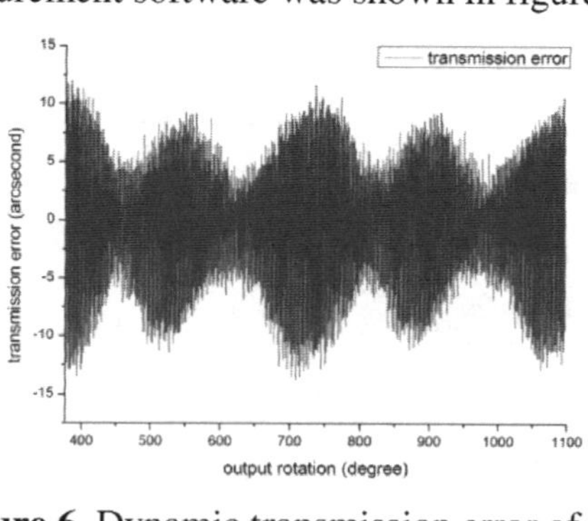

Figure 6. Dynamic transmission error of harmonic reducer

IOP Conf. Series: Materials Science and Engineering **627** (2019) 012013 doi:10.1088/1757-899X/627/1/012013

Based on the above control principle and measurement software, the automatic measurement of transmission performance could be realized.

4. Experiment

In order to verify the feasibility of the measurement system, a harmonic reducer with gear ratio 120 was measured without load, and the measuring speed is 90RPM. The dynamic transmission error curve was given in figure 6. As shown in figure 6, the magnitude of transmission error is ±12.5 arc-second. Therefore, the automatic measurement of transmission performance could be realized by this measurement system. Only when the precise dynamic transmission error curve is obtained, it is possible to separate single transmission error, which is also the direction of further research.

5. Conclusion

In this research, the principle and method for measuring transmission error and mechanical efficiency of harmonic reducer were introduced, and measuring platform was established. The principal factors which may affect the measuring accuracy of this platform were also analysed. Furthermore, the control system consists of a motor driver module, loading module, data acquisition and processing module was introduced, and the measuring software was designed based on the control principle. Finally, a harmonic reducer with a gear ratio of 120 was measured by this system. The result indicates that the measurement system developed in this paper could achieve precision automatic measurement for transmission performance of harmonic reducer.

6. References

[1] Yunwen S and Qingtai Y 1985 *The theory and design of harmonic drive (in chinese)* (Beijing: China Machine Press)

[2] Emelyanov A F 1983 Calculation of the kinematic error of a harmonic gear transmission taking into account the compliance of elements. *Soviet Engineering Research.* 7-10

[3] Xiefeng X, Xingtao D and Linzhi S 1996 Analysis of Transmission Error of Harmonic Gear Drive. *Journal of Zhejiang University of technology.*53-60

[4] Shuvalov S A 1979 Calculation of forces acting on members of a harmonic gear drive. *Russian Engineering Journal.* 5-9

[5] Jianchu Y, Yibao C, Ji Z and Jun Y 2001 Study on a method for calculating gearing meshing efficiency. *Chinese Journal of Mechanical Engineering.* 18-21

[6] Yanabe S , Ito A , Okamoto A , Yamaguchi T and Fujita H 1990 Rotational transmission error of harmonic drive device. *Transactions of the Japan Society of Mechanical Engineers.* 148-153

[7] Weiyong L 2011 Design and implementation of a universal test platform for the transmission accuracy of harmonic reducer (Harbin Institute of Technology) chapter 2 pp 11-22

[8] Hejny S W 1997 Design of a harmonic drive test apparatus for data acquisition and control (Rice University) chapter 4 33-46

[9] Jian-Hong R, Tie-Cai L and Jian S 2014 A Control System Design of Testing Harmonic Drive Reducer Transmission Efficiency. *Modular Machine Tool & Automatic Manufacturing Technique.* 55-57

[10] Zhen C, XinSheng L, ZhiYuan L and Yaqin W 2008 Creeper Gear Mechanical Efficiency Testing System Based on Virtual Instrument. *Instrument Technique & Sensor.* 31-33

[11] Qi L, Xinhui Wand Weidong H 2016 Research of the Comprehensive Performance Test System of the RV Reducer Used in Robot. *Journal of Mechanical Transmission.* 1-3

[12] Hongchun W 2005 Research on automatic testing system of harmonic reducer (Harbin Institute of Technology) chapter 2 pp5-20

[13] Xuesong M, Tao T, Muxun X, Fei Z, Yun Z and Xinmu Y 2009 Dynamic performance comprehensive detection system of harmonic speed reducer. CN101587016

Acknowledgments

This research work was support the National Natural Science Foundation of China (U1508211).

Error Analysis of a Kind of 3D Laser Ball Bar

Liang Xu[1], Yuchen Tian[1], Zhifeng Lou[1*], Kuangchao Fan[1, 2], Liding Wang[1], Xiaodong Wang[1] and Tongqun Ren[1]

[1] School of Mechanical Engineering, Dalian University of Technology, Dalian 116023, China
[2] Department of Mechanical Engineering, National Taiwan University, Taipei 10617, Taiwan, China

Corresponding author: louzf@dlut.edu.cn

Abstract. Laser tracker instruments are used wildly in the measurement of spatial position, but the instruments have complicated structure and high manufacture cost. In this paper, a kind of 3D laser ball bar instrument for measuring spatial position was introduced, of which measuring principle is a passive laser tracking method. Based on the analysis of error sources, the compensation model of 3D laser ball bar was established according to multi-body system error modelling method. Then, a series of calibration methods for measurement errors are proposed, such as the angular error of two rotary stages, transit tilt and laser axis tilt. Measurement experiments indicate that the positioning error of X-axis is reduced from 20μm to 8μm, and the positioning error of Z-axis is reduced from 61μm to 28μm after error compensation.

1. Introduction

Laser Tracker was invented by Lau at the National Institute of Standards and Technology (NIST) in the mid-1980s [1-2]. Many scholars analyzed the measurement accuracies of laser trackers. Lau et al. carried out the early work of error analysis of laser tracker, and studied the influence of the axes' non-orthogonality and beam misalignments on measurement accuracy [2]. Loser and Muralikrishnan analyzed the error sources and proposed a geometrical error model for laser trackers with and without beam steering mirrors respectively [3-4]. Conte establishes a kinematic model of the laser tracker based on D-H method [5]. In addition, some scholars researched on the angular error, multi-station measurement method of laser tracker [6-7], etc.

Nowadays, laser tracking measurement has been used in precision manufacturing and instrument calibration. However, laser trackers have complicated structures and high manufacture cost, and it is difficult to be used in industry wildly. In this paper, a kind of 3D laser ball bar instrument for measuring spatial coordinate was introduced, of which measuring principle is a kind of passive laser tracking method, and structure is very simple.

2. 3D laser ball bar

2.1. Establishment of coordinate systems

As shown in figure 1, the 3D laser ball bar system consists of a two-dimensional rotary stage and an extendable sliding guide that passes through the intersection of two rotational axes. A standard ball is fixed on the end of the guide, which can be held by the motion target and move with it. The system uses a laser Doppler displacement meter (LICS100) for measuring displacement of the extendable sliding guide, and two encoders for measuring rotatory angle of the stages, and then a 3D spherical coordinate system is established.

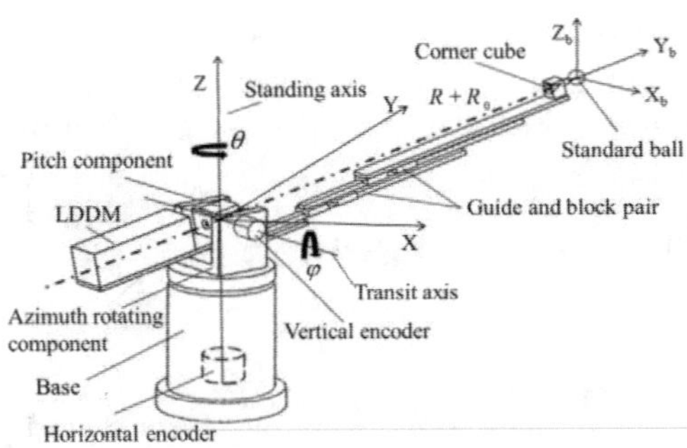

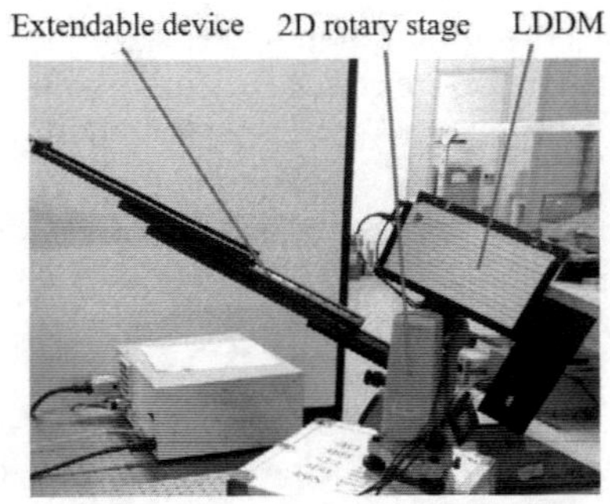

Figure 1. 3D Laser ball bar

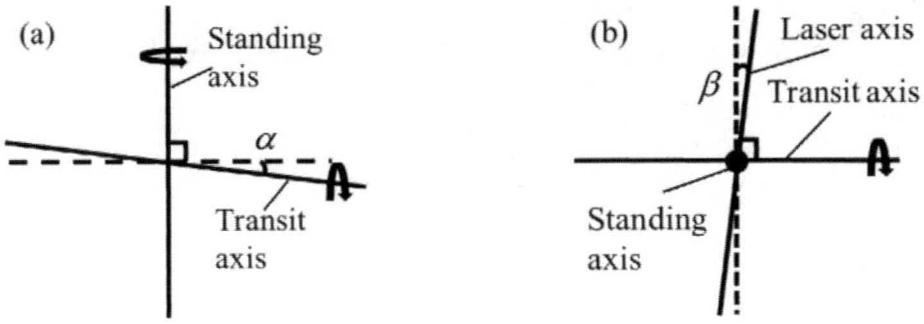

Figure 2. The main geometric error in 3D laser ball bar (**a**) Transit axis tilt; (**b**) Laser axis tilt

2.2. The main error of the 3D laser ball bar

As shown in figure 2, the non-orthogonality between the standing axis and transit axis of the two rotary stages is defined as transit tilt (α). The non-orthogonality between the transit axis and laser axis is defined as laser axis tilt (β), the straightness errors of the extendable guide are defined as δ_x and δ_z.

2.3. Error compensation model of 3D laser ball bar

According to the kinematic relationship of 3D laser ball bar in figure 3, the measurement of the spatial position can be described as follows. The moving target holds the standard ball to move, and extendable device makes a radial translation R along the pitch component (Y_2). Then the pitch component rotates φ around the transit axis (X_1). The azimuth rotating component rotate θ around the base (Z_0). Therefore, the actual coordinate of the standard ball is expressed as Eqs.1 [8].

$$\begin{cases} X_b = -(R_0 + R)(\cos\varphi\sin\theta + \beta\cos\theta - \alpha\sin\varphi\cos\theta) + \delta x\cos\theta + \delta z\sin\varphi\cos\theta \\ Y_b = (R_0 + R)(\cos\varphi\cos\theta - \beta\sin\theta + \alpha\sin\varphi\sin\theta) + \delta x\sin\theta - \delta z\sin\varphi\cos\theta \\ Z_b = (R_0 + R)\sin\varphi + \delta z\cos\varphi \end{cases} \tag{1}$$

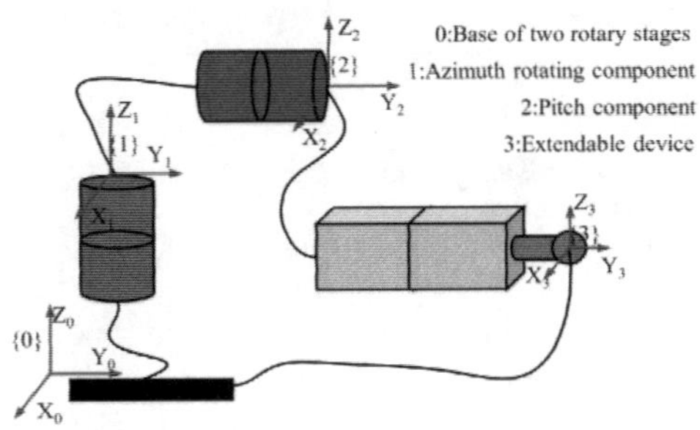

Figure 3. Kinematic relationship of 3D laser ball bar

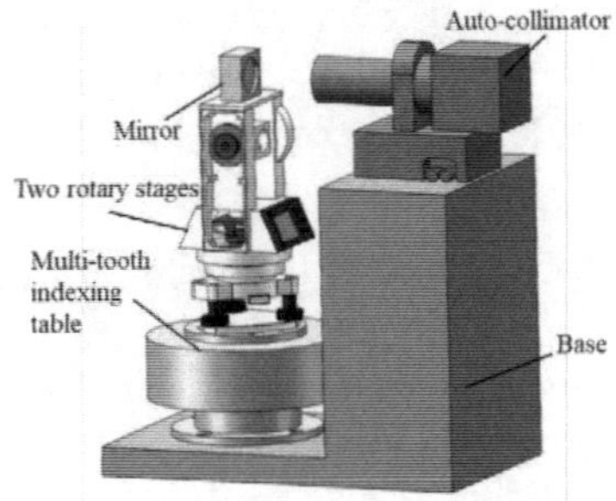

Figure 4. Angular error calibration

3. Errors' measurement and analysis

3.1 Calibration of angular error of two rotary stages

The angular error calibration system is shown as figure 4. When the rotary stage rotates an angle, the multi-tooth indexing reverses the same angle, and then the angular error of the axes will be displayed in the autocollimator. Finally, discrete data is fitted by Eqs.2 based on the harmonic fitting methods [9]. After compensation, the angular error is less than ±1″.

$$\varepsilon(\theta) = c_0 + \sum_{i=1}^{m} [a_i \cos(i\theta) + b_i \sin(i\theta)] \tag{2}$$

3.2 Calibration of base length R_0

The base length is the distance between the standard ball and the intersection of the standing axis and the transit axis when the extendable guide is at initial position. As shown in figure 5, the standard ball is in different positions, and related parameters are obtained. The base length R_0 can be computed according to equation (3).

$$(R_0 + R_1)^2 + (R_0 + R_2)^2 - 2(R_0 + R_1)(R_0 + R_2)\cos\theta_2 = H_2^{\,2} \tag{3}$$

In Eqs.3, R_1 and R_2 are extending displacement from initial position when the standard ball is in position A and position B respectively.

Measurement experiment of the base length is shown in figure 6. The laser ball bar was fixed on a three-axis machine's table, and the standard ball was held by the tool holder. Firstly, the position of the laser ball bar was adjusted in order that the indication of the vertical encoder did not change when the precision three-axis machine tool moved along the X-axis. Then, the displacement in X direction, rotation angle of standing axis, and measurement value of the LDDM were measured. Finally, R_0 can be obtained according to Eqs.3.

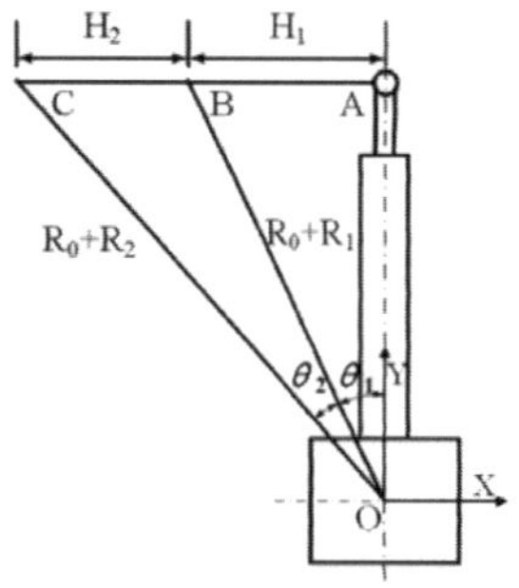

Figure 5. Base length calibration schematic

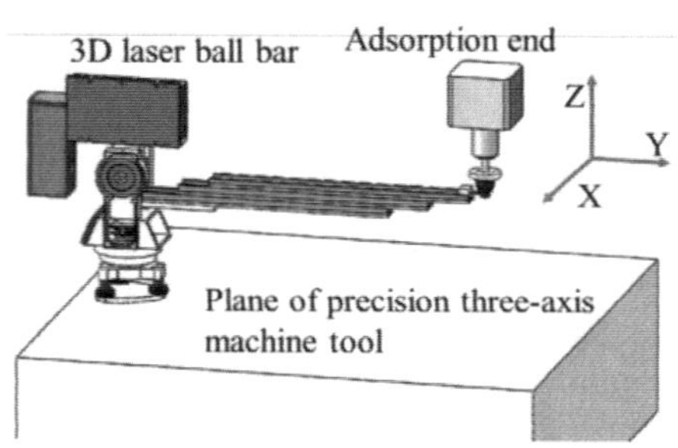

Figure 6. Calibration and comparison system

3.3 Calibration of the tilt errors

If the transit tilt error and the laser axis tilt error exist, the measurement data of horizontal encoder will change continuously when the standard ball moves along Z direction as shown in figure 6. The change of horizontal angle $\Delta\theta$ can be expressed as Eqs.4, where α and β are the transit tilt angle and the laser axis tilt angle respectively, φ is the reading of vertical encoder.

$$\Delta\theta = \alpha \tan\varphi - \beta(\sec\varphi - 1) \tag{4}$$

As shown in figure 6, it is necessary to adjust the base of the 3D laser ball bar to ensure that the reading of the vertical encoder remains unchanged when the 3D laser ball bar moves along the X axis of the machine tool. Then, when the 3D laser ball bar moves along the Z-axis of the machine tool, the change of horizontal angle at different vertical angles can be obtained. After data acquisition, a least-squares fit method is performed to analyze the transit tilt error and the laser axis tilt error according to Eqs.4.

3.4 Straightness measurement of the guide

The straightness errors of the extendable guide were measured by a collimator system [10], and are shown in figure 7.

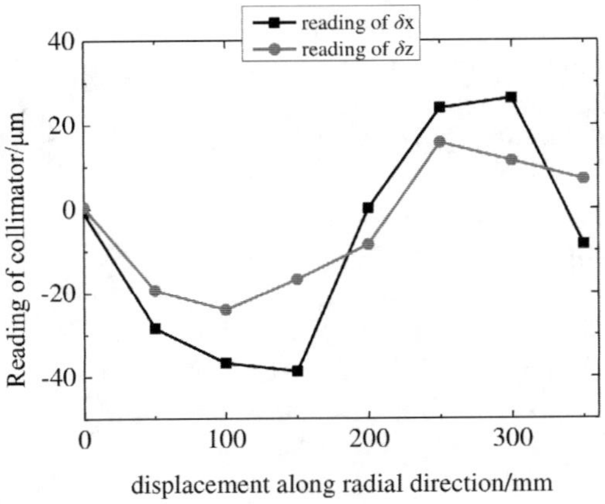

Figure 7. Straightness measured in horizontal plane

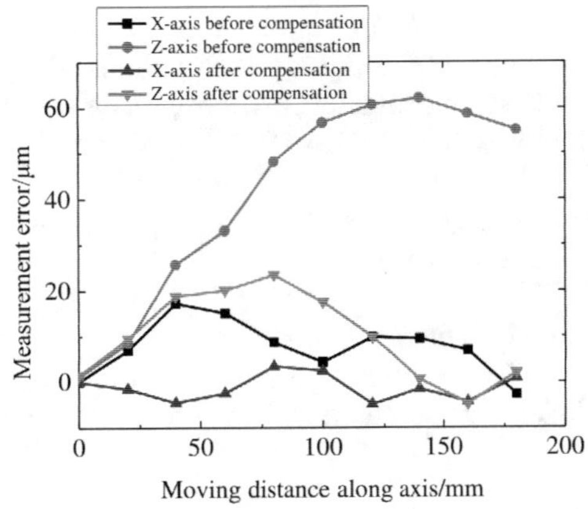

Figure 8. Comparison of positioning error

3.5 Parameters in the 3D ball bar

After measurement experiments, the main parameters can be obtained and are shown in table 1.

Table 1. the main parameter in the 3D laser ball bar.

Parameter	Description	Value
R_0	Base length	567.408 mm
α	Transit axis tilt	108.6 arcsec
β	Laser axis tilt	-5.6 arcsec

4. Calibration experiments

Calibration experiments were conducted on a precision three-axis machine tools, and the coordinate transformation between the machine and the ball bar was carried out firstly. After error compensation, the measurement errors of the ball bar in X direction of the machine were reduced from 20μm to 8μm, and measurement errors in Z direction were reduced from 61μm to 28μm, as shown in figure 8.

5. Conclusion

In this paper, a kind of 3D laser ball bar for measuring spatial coordinate was proposed, and error compensation model and the main errors' measurement methods were developed. Calibration with a precision machine tools indicated that measurement errors of the laser ball bar were reduced greatly.

6. Reference

[1] Muralikrishnan B, Phillips S and Sawyer D 2016 Laser trackers for large-scale dimensional metrology: A review *Precision Engineering* **44** 13-28.

[2] Lau K, Hocken R J, and Haight W C. 1986 Automatic laser tracking interferometer system for robot metrology *Precision Engineering* **8** 3-8.

[3] Loser R and Kyle S 1999 Alignment and field check procedures for the Leica Laser Tracker LTD 500 *Boeing Large Scale Optical Metrology Seminar.*

[4] Muralikrishnan B, Sawyer D, Blackburn C, Phillips S, Borchardt B and Estler W T 2009 ASME B89. 4.19 performance evaluation tests and geometric misalignments in laser trackers *J. Research of the National Institute of Standards and Technology* **114** 21.

[5] Conte J, Santolaria J, Majarena A C, and Aguado S 2015 Laser tracker kinematic error model formulation and subsequent verification under real working conditions *Procedia engineering* **132** 788-795.

[6] Muralikrishnan B, Blackburn C, Sawyer D, Phillips S and Bridges R 2010 Measuring scale errors in a laser tracker's horizontal angle encoder through simple length measurement and two-face system tests *J. Research of the National Institute of Standards and Technology* **115** 291.

[7] Zhuang H, Motaghedi S H, Roth Z S and Bai Y 2003 Calibration of multi-beam laser tracking systems *Robotics and Computer-Integrated Manufacturing* **19** 301-314.

[8] Su S 2002 Study on the methods of precision modeling and error compensation for multi-axis CNC machine tools *National University of Defense Technology*.

[9] Lu R, Li W, Lao D, Zhou W and Ding G 2014 Angular error compensation for laser tracker *Optics & Precision Engineering* **22** 2299-305.

[10] Feng Q, Zhang B and Kuang C 2004 A straightness measurement system using a single-mode fiber-coupled laser module *Optics & Laser Technology* **36** 279-283

Acknowledgments

This project is supported by National Natural Science Foundation of China (U1508211, 51621064), National Key Research and Development Program of China (2017YFF0204800).

Water absorption and thermal degradation behavior of graphene reinforced poly(lactic) acid nanocomposite

OT Adesina [1], T Jamiru [1]. ER Sadiku [2]. OF Ogunbiyi [1] and TA Adegbola [1]

[1]Department of Mechanical Engineering, Mechatronics and Industrial Design, Tshwane University of Technology, Pretoria, South Africa
[2]Institute of NanoEngineering Research (INER) and Department of Chemical Metallurgy and Materials Engineering, Tshwane University of Technology Pretoria,
South Africa
Corresponding author*. Tel (+27) 651073608;
E-mail: oluwagbengaadesina@yahoo.com

Abstract
Thermal stability and high moisture absorption behaviour are major limitations of extensive application of the biodegradable poly(lactic) acid in a variety of technical application which includes the automotive sector. The use of nano fillers and appropriate processing techniques is being considered as a possible route to improve the performance of poly (lactic acid) composites. This study presents the improved performance of PLA composite via the novel processing technique of SPS. The response of the developed composite subjected to moisture absorption over a time range of 6 hrs, 12 hrs and 36 hrs indicated a negligible water uptake. Moreso, the thermogravimetric analysis (TGA) test on the composite reveals thermal stability from room temperature up until 400° C for all the samples. The thermal degradation of the composite was observed above 400°C, indicating a substantial enhancement in its thermal property.

Keywords: polylactic acid, graphene, nanocomposite, water absorption test, Thermogravimetric Analysis.

1. Introduction

Compostable and biodegradable polymers have attracted major research interest due to their potential to mitigate environmental burden. There are several ongoing attempts to replace non-degradable polymers with new sets of biodegradable ones [1,2]. Poly(lactic) acid is considered as the most promising candidate among the biodegradable polymers. Several industries such as packaging, pharmaceutical, textile, automotive, biomedical and tissue engineering are driving towards harnessing the inherent potentials of this eco-friendly polymer. It is a transparent, compostable thermoplastic made from renewable agricultural resources [3]. PLA is regarded as a green polymer because of its ability to degrade into CO_2 and H_2O [4]. It is characterized by high strength, stiffness and easy processing [5]. In spite of the numerous advantages of PLA, few of its drawback are poor water vapour/gas barrier properties and undesirable thermal stability which serve as a restriction for its extended application as bioplastics [6]. Among many approaches adopted by researchers to overcome these limitations are the development of composites, polymer bends and nanocomposites with high aspect ratio organic and inorganic fillers [7,8].

The outstanding mechanical, electrical, thermal and optical features of graphene has made it a focus for research interest in recent times [9,10]. It has shown its great potential as nano-sized fillers in polymer nanocomposite. The use of graphene in polymer material development is often geared towards the property modification of polymer nanocomposite. A significant improvement in material property can be observed even at very low concentration of graphene in a polymer matrix. This can be attributed to the inherent features in graphene. This can be observed when the property of a pristine polymer is being compared with that of a polymer nanocomposite having graphene has nano-filler. Moreso, it has also shown better mechanical and electrical properties when compared with nano clay and other carbon filler polymer nanocomposites [11,12].

The structural application of carbon-based polymer composites exposes it to a variety of environmental conditions which includes high temperature and humidity. The influence of water absorption on the physio- chemical properties of polymer composite has been examined by many researchers. Results obtained revealed that hydrothermal ageing alters the physical and mechanical properties of composite impacting on the performance of the composite in service. In addition, the absorption of water in polymer matrix composite may lead to the plasticization of the polymeric matrix with a resultant significant reduction in the glass- transition temperature and concurrent swelling [13,14]. Moreover, incorporation of fillers into polymeric matrices leads to changes in their glass transition temperature Tg, which can be attributed to the immobilisation of polymer segments close to the surface of the filler particles [15].

The factors influencing moisture absorption in composite materials are temperature, volume fraction of the filler, orientation of reinforcement, nature of filler/ reinforcement; which implies its permeability, polarity and density, area of surface exposed, diffusivity and surface protection [16]. Prolongo et al. [17], studied the water uptake of epoxy composites reinforced with carbon nanofillers. It was reported that carbon nanotubes can significantly decrease the water absorption and water diffusion coefficient by 20-30%. The maximum water content absorbed significantly decreases with the addition of nanofillers, especially in epoxy/CNT composites. The hydrophobic nature of carbon nanofillers is said to be responsible for this regardless of the hydrophilic behaviour of epoxy resin. The decrease of maximum percentage of absorbed water reaches up to 23% for epoxy composite reinforced with 0.1 wt% CNT. Chaharmahali et al. [11] worked on the effect of nano-graphene on the physico-mechanical properties of bagasse/polypropylene composites. The water absorption, thickness swelling, thermal and mechanical properties were evaluated. Results on water uptake claimed a decreased after the addition of nanographene. This was attributed to the fact that graphene layers provided a tortuous path and increased barrier property for water transport. The hydrophobic and water-repelling nature of nano graphene (NG) surface tends to immobilize some of the moisture and inhibits the water permeation in the polymer matrix. This two mechanism suggest that the water absorption has occurred on the surface layer.

This study explores the water uptake behaviour of poly(lactic) acid nanocomposite reinforced with graphene nano particle. It also considers the thermal stability property of the developed material. A very low percentage composition of graphene was added to the poly(lactic) acid matrix. A novel fabrication technique of spark plasma sintering (SPS) was employed in the developed of the nanocomposite. The samples were developed under varied process operating parameters of SPS.

2. Experimental

2.1 Materials

The PLA powder supplied by Micro Powders, Inc., NY was used for the study. According to the product description, a density of 1.23–1.25 g/cm^3 was indicated by the supplier with a maximum particle size of 180 μm and a mesh size of 80. The melting point of PLA of 150–160 °C given by the supplier. Graphene nanoplatelets (GNP) was supplied by Sigma–Aldrich. The GNP used in this experiment has an average diameter of 5μm, an average thickness of 15 nm, and a typical surface area of 50–80 m^2/g. All materials were used as received without any further purification or modification.

IOP Conf. Series: Materials Science and Engineering **627** (2019) 012015 doi:10.1088/1757-899X/627/1/012015

2.2 Fabrication of the Nanocomposite

The procedure applied in the fabrication of these nanocomposites is extensively explained in previous work on optimisation of SPS processing parameters on the density and hardness property of graphene-reinforced poly (lactic) acid nanocomposite [18]. Five percent weighted percentage of GNP was used as reinforcement in the poly(lactic) acid matrix. Design of experiment (DOE) was employed to know the numbers of feasible experimental runs. The samples were prepared through the process of spark plasma sintering at varied conditions of temperature and pressure. A specimen of 10mm thick and 30mm diameter is produced from the admixed powder at full densification by HPD 25 spark plasma sintering (SPS) machine from FCT Systeme GmbH (Raustein, Germany). The heating process ranges from room temperature to the desired temperature with a heating rate of 20 °C /min and holding time of 10 min was considered for this process. The density values were obtained from our previous study and the process parameters which produced each sample were denoted in the alphabetical code as shown in the tables.

2.3. Characterisation

2.3.1 Water Absorption.

The extent of water absorption was examined by complete immersion of the sintered samples in water at room temperature. The initial weight of the samples was recorded before immersion. The immersed composites were taken out periodically from water at a constant time interval of 6 hrs, 12 hrs, 24 hrs and 36 hrs. The water on the surface of the samples was wiped out using a tissue to know the amount of water absorbed in the composite. The amount of water absorbed in the composite was calculated by the weight difference between the samples exposed to water and the dried samples. The weight was measured using digital balance having a precision of 0.01g. The water absorption was carried out using the ASTM D 570-98 (ASTM 1998). The percentage increase in weight was calculated using

$$\textbf{Water Absorption \%} = \{(W_f - W_i) / W_i\} \times 100\ \%$$

Where W_i is the initial weight before immersion and W_f is the final weight after immersion.

2.3.2 Thermal Analysis.

The thermal stability of the nanocomposites was carried out on the GNP/PLA samples to determine the weight loss as a function of temperature. The thermal property was determined using a thermo-gravimetric analyser (TGA; TA Instrument, Model Q500, USA). A constant heating rate of 10° C/min from room temperature to 700°C was maintained under air flow.

3. Results and Discussion

3.1 Thermal stability

The TGA weight loss of the nanocomposite is shown in Figure 1. The processing technique of SPS was influential to the development of highly densified nanocomposite sample. However, there was no wide variation in the thermal stability from the room temperature to 400°C as a result of varying operating parameters (temperature and pressure) of SPS. The temperature values obtained were within close range. Although there exists a difference in the residual remaining as shown in Figure 1. A correlation was observed between the densification of the nanocomposite samples and the amount of residuals. The sample with the greatest densification has the highest residue above 400°C. The level of thermal stability witnessed could be associated with the shielding effect due to the present of flake-like nanofiller and obstructed diffusion of volatile decomposition product [19,20].

Table 1: The temperature at 5% weight loss (T5%), 50% weight loss (T50%), 75% weight loss (T75%) and the percentage of char residue at 400° C.

Sample code	$T_{5\%}$ (° C)	T50% (° C)	T75% (° C)	Residual at 400 ° C	Density
A	344.93	376.13	385.31	2.261655	1.276
B	346.30	377.93	388.26	4.666572.	1.285
C	347.54	381.33	391.79	4.963210	1.288
E	349.12	379.71	390.13	4.731041	1.287
F	347.01	376.30	385.87	2.844818	1.278
G	352.15	377.92	387.26	3.34310	1.283

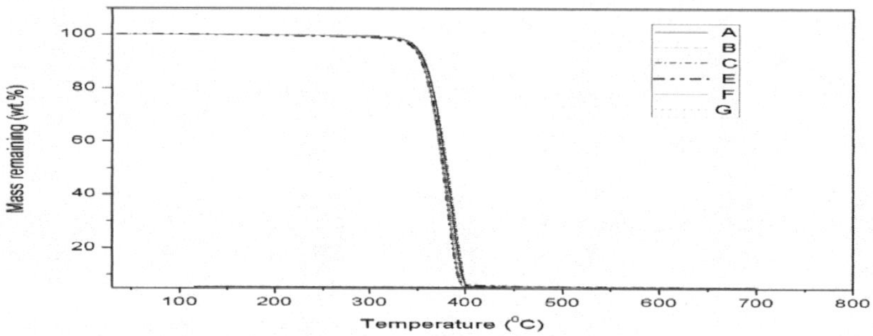

Figure 1: **TGA of Sintered GNP/PLA Nanocomposite samples under varying process conditions.**

Tables 2, 3 and 4 present the influence of graphene nanoparticle (GNP) and more importantly manufacturing process on the water absorption characteristics of GNP/PLA nanocomposite. Throughout the immersion time, a very insignificant amount of water was absorbed by the nanocomposite. This is due to the hydrophobic nature of GNP and the densification property of the sintered compact which recorded above 99% densification and minimal porosity on the developed sample. The synergetic effect brought about by the inclusion of GNP and the densified sintered compact was able to modify the hydrophilic nature of polylactic acid in the composite. A well-compacted sample of stronger adhesion between the reinforcement and the matrix was produced while the hydrophobic nature of graphene nanoplatelets surface immobilise some of the moisture, slowing down the rate of water absorption. A similar observation was found with [11] as graphene inclusion also reduced the water absorption in the developed nanocomposite immobilising some of the moisture thereby inhibiting water permeation into the polymer matrix.

Table 2: Water absorption test results for 6 hrs.

Sample code	Dry wgt. (gm)	Wgt. gained (6hrs)	Change in wgt. (gm)	% gained in wgt.	Density
A	4.31	4.32	0.010	0.2304	1.276
B	3.98	3.99	0.010	0.0025	1.285
C	4.32	4.32	0.000	0.0000	1.288
E	4.23	4.23	0.000	0.0000	1.287
F	3.71	3.81	0.100	0.0270	1.278
G	4.07	4.09	0.020	0.0049	1.283

Table 3: Water absorption test results for 12 hrs

Sample code	Dry wgt. (gm)	Wgt. gained (12hrs)	Change in wgt. (gm)	% gained in weight
A	4.31	4.33	0.02	0.00464
B	3.98	4.00	0.02	0.00503
C	4.32	4.32	0.00	0.00000
E	4.23	4.24	0.01	0.00236
F	3.71	3.81	0.01	0.02695
G	4.07	4.10	0.03	0.00700

Table 4: Water absorption test results for 24 hrs.

Sample Code	Dry wgt.(gm)	Wgt. gained (24hrs)	Change in wgt.(gm)	% gained in weight
A	4.31	4.34	0.03	0.00696
B	3.98	4.00	0.02	0.005025
C	4.32	4.33	0.01	0.002315
E	4.23	4.24	0.01	0.002364
F	3.71	3.81	0.10	0.026954
G	4.07	4.10	0.03	0.007371

4. Conclusion.

The hydrophobic nature of the nano filler, GNP in the developed PLA nanocomposite was greatly significant as it alters the hydrophilic nature/water absorbing property of polylactic (PLA) in the matrix. Thereby, resulting in an insignificant amount of water uptake in the nanocomposite. Moreso, the influence of the fabricating process in producing a highly densified nanocomposite with negligible porosity cannot be over emphasized. This also contributed to the reduction in water absorption rate of the nanocomposite. The presence of GNP as reinforcement in the nanocomposite is also responsible for the improvement in the thermal stability of the composite at a uniform steady state of around 400° C.

5. References

[1] Huneault MA and Li H 2012. Preparation and properties of extruded thermoplastic starch/polymer blends. *Journal of Applied Polymer Science*, *126*(S1), pp.E96-E108.

[2] Pilla S, Kim SG, Auer GK, Gong S and Park, C.B., 2009. Microcellular extrusion-foaming of polylactide with chain-extender. *Polymer Engineering & Science*, *49*(8), pp.1653-1660.

[3] Najafi N, Heuzey MC and Carreau PJ 2012. Polylactide (PLA)-clay nanocomposites prepared by melt compounding in the presence of a chain extender. *Composites Science and Technology*, *72*(5), pp.608-615.

[4] Liu R, Luo S, Cao J and Peng Y 2013. Characterization of organo-montmorillonite (OMMT) modified wood flour and properties of its composites with poly (lactic acid). *Composites Part A: Applied Science and Manufacturing*, *51*, pp.33-42.

[5] Huda MS, Drzal LT, Mohanty, A.K and Misra M 2008. Effect of fiber surface-treatments on the properties of laminated biocomposites from poly (lactic acid) (PLA) and kenaf fibers. *Composites science and technology*, *68*(2), pp.424-432.

[6] Oksman K, Mathew AP, Bondeson D and Kvien I 2006. Manufacturing process of cellulose whiskers/polylactic acid nanocomposites. *Composites science and technology*, *66*(15), pp.2776-2784

[7] Sanchez-Garcia MD and Lagaron JM 2010. On the use of plant cellulose nanowhiskers to enhance the barrier properties of polylactic acid. *Cellulose*, *17*(5), pp.987-1004.

[8] Petersson L, Kvien I and Oksman K 2007. Structure and thermal properties of poly (lactic acid)/cellulose whiskers nanocomposite materials. *Composites Science and Technology*, *67*(11- 12), pp.2535-2544.

[9] Cardinali M, Valentini L, Kennya JM and Mutlay I 2012 Graphene based composites prepared through exfoliation of graphite platelets in methyl methacrylate/poly(methylmethacrylate). *Polym Int* 61:1079–1083

[10] Sheshmani S and Amini R 2013 Preparation and characterization of graphene based nanocomposite materials. *Carbohydr Polym* 95:348–359

[11] Chaharmahali, M., Hamzeh, Y., Ebrahimi, G., Ashori, A. and Ghasemi, I., 2014. Effects of nano-graphene on the physico-mechanical properties of bagasse/polypropylene composites. *Polyme bulletin*, *71*(2), pp.337-349.

[12] Kuilla T, Bhadra S, Yao D, Kim NH, Bose S, Lee JH (2010) Recent advances in graphene based polymer composites. *Prog Polym Sci* 35:1350–1375.

[13] Maggana, C and Pissis, P., 1999. Water sorption and diffusion studies in an epoxy resin system. *Journal of Polymer Science Part B: Polymer Physics*, *37*(11), pp.1165-1182.

[14] Kotrotsos, A., Vavouliotis, A., Tsantzalis, S. and Kostopoulos, V., 2014. Effect of CNT modified matrix of epoxy CFRPs on hydrothermal behaviour of material. Evaluation of water uptake using electrical resistance measurements. *Plastics, Rubber and Composites*, *43*(4), pp.122-129. [15] Zhou J and Lucas JP

1999. Hygrothermal effects of epoxy resin. Part II: variations of glass transition temperature. *Polymer*, *40*(20), pp.5513-5522.

[16] Lin YC, Hsu FH and Wu TM 2013. Enhanced conductivity and thermal stability of conductive polyaniline/graphene composite synthesized by in situ chemical oxidation polymerization with sodium dodecyl sulfate. *Synthetic metals*, *184*, pp.29-34.

[17] Prolongo SG, Gude, M.R. and Urena A 2012. Water uptake of epoxy composites reinforced with carbon nanofillers. *Composites Part A: Applied Science and Manufacturing*, *43*(12), pp.2169-2175.

[18] Adesina, OT, Sadiku, ER, Jamiru T, Ogunbiyi OF., Beneke LW and Adegbola AT 2019. Optimization of SPS processing parameters on the density and hardness properties of graphene reinforced polylactic acid nanocomposite. *The International Journal of Advanced Manufacturing Technology*, pp.1-12

[19] Murariu, M., Dechief, A.L., Bonnaud, L., Gallos, A., Fontaine, G., Bourbigot, S. and Dubois, P., 2010. The production and properties of polylactide composites filled with expanded graphite. *Polymer Degradation and Stability*, *95*(5), pp.889-900.

[20] Yang, X., Li, L., Shang, S. and Tao, X.M., 2010. Synthesis and characterization of layer-aligned poly (vinyl alcohol)/graphene nanocomposites. *Polymer*, *51*(15), pp.3431-3435.

Acknowledgement
The authors would like to thank the Tshwane University of Technology Pretoria for providing the resource to carry out this research work.

IOP Conf. Series: Materials Science and Engineering **627** (2019) 012016 doi:10.1088/1757-899X/627/1/012016

Evaluation and optimization of machining performance for rotary ultrasonic machining of Al/SiC composite material

Jatinder Kumar

Department of Mechanical Engineering, NIT Kurukshetra, India

jatin.tiet@gmail.com

Abstract. In manufacturing industries, Al/SiC composite material is highly demanded due to its excellent properties. However, the difficulties in its machining restrict the application and competitiveness of this material. This investigation is aimed at studying the impact of different experimental conditions (feed rate, spindle speed, and ultrasonic power) on responses of interest (material removal rate, surface roughness, and chipping thickness) in rotary ultrasonic machining of Al/SiC composite material. The experiments have been planned by using full factorial design. Analysis of variance (ANOVA) is also employed to identify the significant factors and the optimized process settings for the different responses. The experimental results showed that feed rate was the most significant factor for material removal rate and surface roughness. However, all the factors were found to be insignificant for chipping thickness. From the surface topography analysis of the machined surface, no surface damage in the form of micro-cracks etc. was visible.

Keywords: Rotary ultrasonic machining, full factorial design, chipping, optimization

1. Introduction

Composites materials offer some special advantages in the form of superior mechanical, thermal and chemical characteteristics along with higher specific strength. Al/SiC is a metal-ceramic composite material that consists of silicon carbide particles distributed in specific aluminum alloy matrix. This material offers a unique combination of high strength to weigh ratio, high value of elastic modulus, good wear resistance and thermal stability, which is generally better than the non-reinforced systems [1]. Hence, this composite is a potential structural material, specifically for applications in aerospace and automotive sectors [1-3]. Also, the Al/SiC composite combines the high thermal conductivity of metal and low coefficient of thermal expansion of ceramic which could be very essential in these applications.

Machining Al-SiC materials with machining techniques based on thermal energy such as, electric discharge machining (EDM) results into cracks, recast layer, and heat affected zone (HAZ) on the machined surface, and variation of mechanical properties [1-3]. These defects further result in reduction of resistance to wear, corrosion and hardness. Consequently, the quality of product is severely affected. For processing Al-SiC composites, rotary ultrasonic machining (RUM) process could be a feasible alternative, as the removal of material is purely mechanical, and also the method is free of all the limitations associated with thermal energy-based machining methods.

The rotary ultrasonic machining was invented by Percy Legge in 1964, at United Kingdom Atomic Energy Authority. This process is a hybrid of conventional grinding and ultrasonic machining as far as the material removal mechanism is concerned. A core drill with diamond abrasive particles bonded

over the metal surface is vibrated at ultrasonic frequency and fed longitudinally, at a static feed rate or a force. The Coolant fluid is pumped through the hole at the core of the drill, which flushes the swarf, cools the machining zone and the prevents interruptions in machining. Rotray USM has been reported to yield much higher machining rates as compared to static USM [4]. Few investigations have been reported on evaluation of the machinability for rotary USM, for different materials ranging from titanium alloys [5], glass [6] and CFRP [7], however, these have been limited to use of either One factor at a time approach (OFAT) or some random experiments for computation of the effects of vital process parameters on the desired response variables. Designed experiments offer some unique advantages such as estimation of interaction effects, segregation of noise, and optimization of response variables under a range of input parameters etc.
From a comprehensive literature review on Rotary USM, it was concluded that a critical need exists to evaluate the machining performance and surface characteristics in rotary ultrasonic machining of Al/SiC composite material, through systematically designed experiments, with an objective to provide the machining community some optimized process conditions/settings in this regard.

2. Materials and methods

Al alloy 6061/10% SiC composite is a metal-ceramic composite material consisting of silicon carbide particles dispersed in a matrix of aluminum alloy 6061. The composition of aluminum alloy is detailed in table 1.

Table 1. Composition of aluminum 6061 alloy

Si	Fe	Cu	Mn	Mg	Cr	Zn	Ti	Other	Al
0.4-0.8	0.0-0.7	0.15-0.40	0.0-0.15	0.8-1.2	0.04-0.35	0.0-0.25	0.0-0.15	0.05-0.15	Balance

Table 2. Process parameters selected for the research

S. No.	Symbol	Parameters	Level 1	Level 2	Level 3
1.	A	Feed rate (mm/sec)	0.0012	0.0080	0.0180
2.	B	Speed (rpm)	3000	3600	4200
3.	C	Ultrasonic power (%)	40	60	-----

Stir casting method was used to fabricate the composite. A controlled argon environment was used while melting the Al alloy, in order to prevent the porosity in the composite samples. The Al alloy stock was heated at 750 °C for 30 minutes, till complete melting. Graphite impeller was used to create the vortex, through stirring at 600 rpm. Specific amount of SiC particles (10% Vol.) was preheated and then poured into the crucible containing Al alloy in molten form. To ensure that the SiC particles get uniformly distributed in the Al alloy matrix, stirring was continued while adding the reinforcement. The casting in a permanent mould was fulfilled through bottom pouring.
Table 2 shows the process parameters and their levels selected for the present study. The levels of input factors were determined on the basis of a pilot experimentation, which was conducted using one factor at a time approach. The experiments were performed on 'Kneemill-10 model' (Sonic-Mill, Albuquerque, USA) rotary ultrasonic machining set-up. The different components of RUM setup such as coolant system, ultrasonic spindle, power supply, and control panel have been shown in Figure 1. Figure 2 shows the diamond drill used for the experiments. Diamond core drills were provided by N.B.R. diamond Tool Corporation (LaGrangeville, NY, USA) .The inner and outer diameters of the hollow tool were 4 mm and 5 mm respectively.

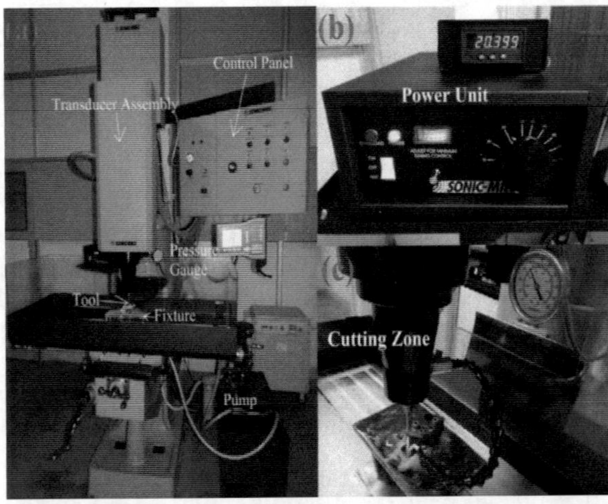

Figure 1. Experimental set-up for RUM

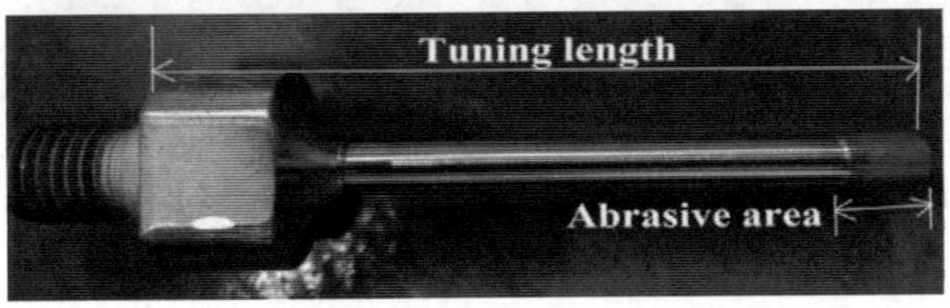

Figure 2. The diamond drill used for machining

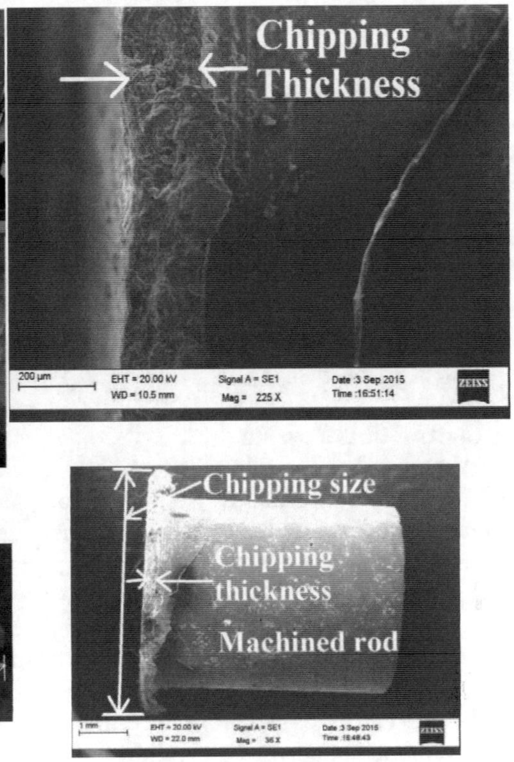

Figure 3. Measurement of chipping thickness by SEM

Table 3. Experimental results

Exp. No.	Feed (mm/sec)	Speed (rpm)	Power (%)	MRR (mm³/sec)	Roughness (µm)	Chipping thickness (mm)
1	0.0012	3000	40	0.058	0.838	0.274
2	0.0012	3000	60	0.063	0.816	0.319
3	0.0012	3600	40	0.061	0.796	0.289
4	0.0012	3600	60	0.067	0.781	0.378
5	0.0012	4200	40	0.064	0.777	0.244
6	0.0012	4200	60	0.073	0.711	0.324
7	0.0080	3000	40	0.074	1.163	0.292
8	0.0080	3000	60	0.084	1.100	0.324
9	0.0080	3600	40	0.078	1.007	0.368
10	0.0080	3600	60	0.090	0.917	0.359
11	0.0080	4200	40	0.085	0.977	0.36
12	0.0080	4200	60	0.093	0.884	0.352
13	0.0180	3000	40	0.139	2.099	0.296
14	0.0180	3000	60	0.148	2.002	0.267
15	0.0180	3600	40	0.148	1.866	0.313
16	0.0180	3600	60	0.161	1.716	0.274
17	0.0180	4200	40	0.151	1.615	0.294
18	0.0180	4200	60	0.170	1.496	0.392

IOP Conf. Series: Materials Science and Engineering **627** (2019) 012016 doi:10.1088/1757-899X/627/1/012016

The mesh size of the abrasive grits bonded over the cutting face was 100. Mobilemet® S122 water-soluble cutting oil (MSC Industrial Supply Co., Melville, NY, USA) was used as the coolant fluid (in 1:15 ratio). MRR was computed by dividing the volume of material removed by the machining duration required to drill a hole. The stop watch was used to record the time taken to drill each hole. Surface roughness (SR) was measured by using a surface roughness tester (Zeiss Surfcom, USA). Chipping thickness was directly observed from SEM images with the help of software (figure 3).

In this study, full factorial design approach is used for conducting the experiments. Use of full factorial approach allows assessing the main effects; Interaction effects with best accuracy, as all the parametric combinations are tested. Two replicates of entire experiment, which consist of 36 experimental runs in total, were run in a completely randomized manner to minimize the effect of noise factors (table 3).

3. Results and discussion

ANOVA test has been performed on the experimental data (detailed in table 3) to assess the significance of the variables. Table 4 depicts the results of ANOVA test for MRR, SR, and CT. For the material removal rate, feed rate is the most significant parameter; other factors such as power rating and spindle speed are also found to be significant. None of the interaction effects (feed rate and spindle speed, feed rate and ultrasonic power; spindle speed and ultrasonic power) have been observed to be significant for MRR. For surface roughness, all the factors were observed as significant however, feed rate is revealed as the most significant parameter. Two interaction effects are also observed to be significant for SR. For chipping thickness, none of the factors or interactions was found to be significant. Three confirmation experiments were performed at the optimized process settings and the results obtained confirmed the validity of predicted optimal values, as both values were found to be in good agreement (table 4).

Table 4. ANOVA results for MRR, SR, and CT

		MRR		**SR**		**CT**		**Optimized values (mean values)**
Source	DOF	*F*	*P*	*F*	*P*	*F*	*P*	For MRR
A	2	2404.40	0.000*	7479.20	0.000*	2.14	0.33	Predicted-0.164 mm^3/sec
B	2	38.05	0.002*	450.54	0.000*	1.75	0.284	Actual- 0.160 mm^3/sec
C	1	85.37	0.001*	124.96	0.000*	2.89	0.165	
A×B	4	2.39	0.210	103.06	0.000*	1.16	0.444	For SR
A×C	2	3.41	0.137	12.57	0.018*	1.58	0.311	Predicted- 0.74 μm
B×C	2	1.12	0.410	1.91	0.271	0.68	0.557	Actual- 0.777 μm
Error	4							
Total	17							For CT
A-Feed rate, B-Spindle speed, C-Power rating, *Significant at 95% confidence level.								Predicted- 0.284 mm Actual- 0.278 mm

Optimized Process Settings	(For MRR)	(For SR)	(For CT)
Feed rate	0.0180 mm/sec	0.0012 mm/sec	0.0012 mm/sec
Spindle speed	4200 rpm	4200 rpm	3000 rpm
Power rating	60%	60%	40%

To assess the effect of machining on the surface, surface topography analysis was performed. The SEM images depicted in figure 5 show no surface damage in the form of micro-cracks or built up layers etc. Though adherence of few micro chips on the surface cannot be ruled out, still the surface created post machining with rotary USM (RUM) appears to be possessing fine surface finish and free of surface irregularities.

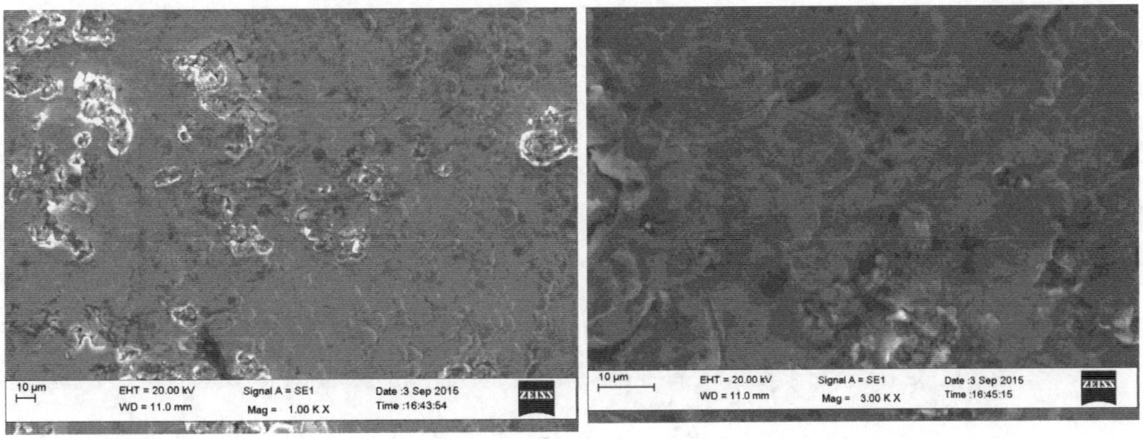

Figure 5. Microstructures of machined surface (a)-exp. 8, (b)- exp.18

4. Conclusions and future scope of work

This experimental research has been focussed on exploring the machinability of Al/Sic Composite material using rotary USM, a hybrid machining method. Following conclusions may be drawn:

1. For MRR and Surface quality, all the input parameters investigated in the study have been found to be significant. Feed rate has been found to cause highest variation in both responses. For surface roughness, two interaction effects have also been found to be significant (feed and spindle speed, feed and power rating).
2. Edge chipping thickness has been effectively controlled through designed experiments. None of the factors or their interactions has been found to be significant for chipping thickness.
3. From the surface topography examination through SEM, no surface damage was apparent for the machined surface, for selected samples. The high quality surface can also be verified from the SEM images.
4. Further investigation may include parameters such as type of bond (for abrasive), diameter of diamond tool, grit size of the abrasive etc., with a view to develop machinability data for RUM.

5. References

[1] Singh, B, Kumar J and Kumar S 2014 *Materials and Manufacturing Processes* ***29*(3)** 287-297.
[2] Singh, B, Kumar J and Kumar S 2015 *Materials and Manufacturing Processes* ***30*(3)** 303-312.
[3] Cong W L, Pei Z J, Deines T W, Srivastava A and Treadwell C 2012 *Ultrasonics* **52(8)** 1030–1037.
[4] Ya G, Quin H W and Yang S C 2002 Journal *of Materials Processing Technology* **129(1/3)** 182–185.
[5] Churi N J, Pei Z J, Treadwell C 2006 *Machining Science and Technology* **10(3)** 301-321.
[6] Lee M H, Sudin I, Ken G E and Zaharim A 2008 *Proc. Int. Conf. Applied Mathematics 214-219.*
[7] Liu J, Zhang D, Quin L and Yan L 2012 *Int. J. Machine Tools and manufacture* **53** 141-150.

Acknowledgments

The author is thankful to N.B.R. diamond Tool Corporation (LaGrangeville, NY, USA*)* for *providing* the diamond drills for this research. Author thanks Mr. Charlie Wilhite (Sonic-Mill, USA) for providing technical support. Author is grateful to NIT Kurukshetra for providing the necessary financial support.

Mechanical performance of GNP/PLA nanocomposite under varied SPS process parameters

OT Adesina [1]. T Jamiru [1]. ER Sadiku [2]. MO Durowoju[3]. OF Ogunbiyi [1]

[1]Department of Mechanical Engineering, Mechatronics and Industrial Design, Tshwane University of Technology, Pretoria South Africa
[2]*Institute of NanoEngineering Research (INER) and Department of Chemical Metallurgy and Materials Engineering, Tshwane University of Technology Pretoria, South Africa*
[3]*Department of Mechanical Engineering, Ladoke Akintola University of Technology, Nigeria.*
Corresponding author*. Tel (+27) 651073608
E-mail: oluwagbengaadesina@yahoo.com

Abstract
The study examines the mechanical performance of GNP/PLA nanocomposite developed under varied SPS process parameters. The samples were fabricated using a novel Spark Plasma Sintering technique. In this research, the effect of sintering process in terms of sintering pressure and temperature has been studied by focusing on the mechanical properties of GNP/PLA nanocomposites. The densification of the samples was obtained via the Archimedes' principle while hardness results was obtained using the micro hardness tester. The results show that the density and hardness increased with increasing sintering temperature and pressure. The samples were prepared under uniaxial pressure of 20 MPa and 30 MPa. The sintering temperature of 135 °C and 160 °C was used in developing the samples. The synergetic influence of these process parameters and graphene inclusion was significant in obtaining above 99% densification and with a concurrent enhancement in the hardness property of the developed composite.

Keywords: polylactic acid, graphene, nanocomposite, densification, hardness

1. Introduction

The ecological and environmental friendliness of bio-based composite has attracted major researchers in recent times. This is in a bid to mitigate the implication of petroleum based material on human, environment and ecological lives. In addition to the need to reduce the over dependency on petroleum resources that is continuously depleting. The use of bio degradable polymer plays an important role in solving public hazards of polymer materials and maintaining ecological balance [1]. However, the application of polylactic acid can be found in packaging, textiles, engineering and biomedical applications, hereby rising to the second place in terms of global demand after starch based plastics in the bio-based plastic category [2].

Poly(lactic) acid is a fully bio-based polymer sourced via the fermentation of agricultural products such as corn, potato and sugar beets. Its basis of attractiveness is own to its biodegradability, recyclability and composability. Its production also consumes carbon dioxide [3,4 &5]. In addition, its requires 25-55% less energy to produce compare to the petroleum based polymers [6]. It is known for its good mechanical properties, excellent barrier capacity and ease of processing in numerous applications. However, it high brittleness, low impact strength, weak thermal stability and water uptake are major drawback for its extended commercial applications [7,8]. The modification and enhancement in the mechanical, physical and thermal properties of poly(lactic) acid is made possible via the use of fibres and fillers as reinforcement in the poly(lactic) acid matrix [9]. The incorporation of nano-sized reinforcement within a polymer matrix produces a nanocomposite material. A polymer nanocomposite refers to multi-phase materials where at least one of the constituent phase which is mostly nano-filler has one dimension in nanoscale range [10,11]. The synergetic effect of reinforcing a biodegradable polymer with nano-fillers compensate for the limitation of the polymer matrix.

The promotion of biopolymer can be viewed in three aspects in terms of its performance, processing and cost. The performance and processing are major issues affecting biodegradable polymers [1]. According to Norazlina et al. [2], the physical, mechanical and barrier properties of PLA depends on the solid state morphology and its crystallinity. The low crystallisation rate and low crystallinity of PLA serves as a major limitation for its extended applications like automotive and electronics [12]. The use of organic or inorganic fillers for reinforcement in polymer matrix serves as heterogeneous nucleating agents to elevate the crystallinity property of PLA [13]. Among the various fillers such as carbon nanotube, clay, graphene(GN) as heterogeneous nucleation is found to be more effective [14,15].

According to the review on the research progress on the preparation of bio-degradation PLA/ Graphene nanocomposites by Li et al. [16], it was inferred that the manufacturing process, technique and appropriate content ratio of graphene played a major influence in improving the properties of the nanocomposites.

Sintering refers to the process of firing and consolidating powders at temperatures lower than their melting point, where diffusional mass transport leads to bonding between particles and the formation of a dense body. Sintering process engages the external pressure and electric field simultaneously to enhance the densification of the powder compact. In SPS, a pulsed direct current is allowed to pass through a conducive die and through the sample. The pulsed direct current produces an electric field during the sintering process [18]. This process is advantageous in producing a highly densified compact within the shortest time, low energy consumption and nano-sized powder can be sintered without considerable grain growth. The stages of SPS occur under applied pressure, increasing temperature and holding time within a short duration of (5-25 min), thereby making control of the microstructure and grain growth achievable mostly in nanostructured materials [17]. The simultaneous application of temperature and pressure in SPS leads to high densification and low porosity can be achieved. This results in outstanding mechanical properties such as strength and hardness [18].

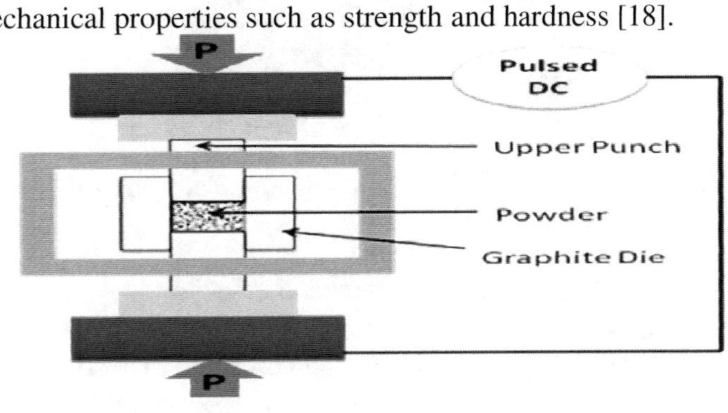

Figure 1: Schematic diagram of spark plasma sintering technique [23]

Several studies on the spark plasma sintering of polymers and polymer composites have emphasised the role of applied pressure and temperature on the mechanical property of sintered compact. Tanaka et al. [19] investigated the wear property of polyimide filled with carbon and diamond particle. The composites were developed using SPS technique. The pressure was set at 50 MPa and a temperature

220°C. At elevated sintering temperature, cracks were noticed and anti- wear properties of the sintered compart deteriorated. Schwertz et al. [20] also worked on the consolidation of polyimide and polyetheretherketone using Spark Plasma sintering technique. Relative densities above 99% was reached for PI at temperature as low as 320 °C and 200 °C for PEEK. The sintered samples show enhancement in the compression strength and hardness property.
This study investigates the influence of process parameters of novel SPS technique on the mechanical performance of graphene reinforced poly(lactic) acid nanocomposite. The parameters of influence in this study are the pressure and temperature while the properties investigated are the density and hardness properties.

2 Experimental Procedure

2.1 Materials
Polylactic acid powder, a Micro Powders, Inc., was used in this study. According to the supplier the melting point was 150-160 °C and the density of 1.23-1.25 g/cm^3. Graphene nanoplatelet from Sigma Aldrish with an average diameter of 5μm and average thickness of 15nm was used. Both materials were used without any further modification or purification.

2.2 Preparation of samples
The sample containing the blend of graphene (GNP) and polylactic acid (PLA) was mixed in a dry environment using Turbula Shaker Mixer T2F for 6hrs for homogeneity and with 5% weight composition of graphene nanoplatelets. The appropriate amount of mixed powders required to produce cylindrical specimen of 10mm thick and 30mm diameter were poured into the graphite die. Graphite sheet were inserted between the graphite die, powders and plungers for easy removal of sintered compact. The admixed powder is then sintered using the Spark Plasma sintering (SPS) machine from FCT Systeme GmbH (Raustein Germany) under varied process parameters. Details can be found in our previous work [21]. GNP/PLA samples were sintered at temperature of 135 °C and 160 °C each under a pressure condition of 20MPa and 30MPa. Figure 2 shows a typical sample of the sintered nanocomposite. This is to investigate the effect of pressure and temperature variation on the density and hardness property of the nanocomposite.

Figure 2: The developed SPSed 30mm diameter and 10 mm thick GNP/ PLA nanocomposites.

2.3 Characterisation
The densities of the sintered samples were determined via the Archimedes principle. The Vickers micro hardness (Hv) at room temperature was measured by the micro hardness tester (Future-tech) at a load (P) 0f 100gf(1.0N) and a dwell time of 15 sec. Arithmetic mean of six successful indentation recorded to ensure accuracy of the data.

3. Experimental results and discussion.
The selection of the temperature values used in this study is based on our previous work [21] where it was observed that below 135 °C, the admixed powders are not consolidated and above 160 °C, the powder melted out of the die in the SPS chamber. In accordance with the information provided by the manufacturer's data sheet, PLA has a melting temperature within the range of (150-160) °C. Also, the sintering of powders must be below the lowest melting point of the powders in the admixture.

Table 1: Density and hardness value of sintered GNP/PLA nanocomposite under different operating parameters

Process Parameters	Notation	Density(g/cm^3)	Hardness(HV)
135 °C, 20 MPa	A	1.278	235.94
135 °C, 30 MPa	B	1.282	243.31
160 °C, 20MPa	C	1.285	257.42
160 °C, 30 MPa	D	1.288	265.00

From the results provided Table 1, for both temperatures (135 and 160) °C, an increase in pressure results in the enhancement in densification which consequently increases the hardness value of the sintered compact. More so, at both pressure conditions, the effect of temperature increase is observed with an increase in both the densification and hardness value of the sintered nanocomposite compact.
The increase in densification is as a result of compressive stress created by the increase pressure which results in improved contacts between the powder particles and a decrease in porosity level due to increase in compaction. Increasing pressure breaks agglomeration, enabling the re arrangement of particles. Thereby removing pores spaces with homogeneous and limited grain growth facilitating increased packing [18,22]. Furthermore, the presence of graphene nanocomposite enhanced the crystallinity property of polylactic acid there improving its hardness property.

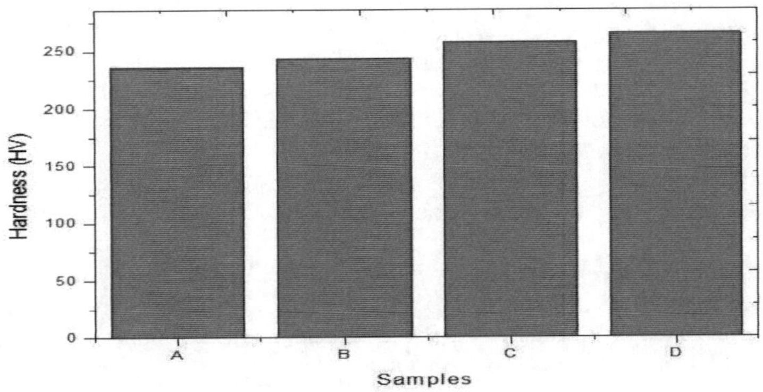

Figure 3: shows the increasing trend of hardness value with respect to varying process parameters.

4. Conclusion

GNP/PLA nanocomposite had been synthesized by Spark plasma sintering technique. The influence of temperature and pressure variation was examined on the density and hardness property of the nanocomposite. The results can be concluded as follows;

1. The density and hardness increases with increase in sintering temperature and pressure within the range of 135-160° C and 20-30 MPa
2. The densification of above 99% were obtained for all the developed nanocomposite which implies a minimal porosity in the material.
3. A developed nanocomposite is of high relevance automobile application based on its light weightiness and eco friendliness. This will help mitigate the rate of carbon emission and other greenhouse gases.
4. The inclusion of graphene nanoplatelets acts as nucleating agent increasing the crystallinity property of the polylactic acid for better mechanical property such as hardness.
5. The process parameters have a major influence on the mechanical performance of the sintered compact.

5. References

[1] Sun, J., Shen, J., Chen, S., Cooper, M., Fu, H., Wu, D. and Yang, Z., 2018. Nanofiller reinforced biodegradable PLA/PHA composites: Current status and future trends. *Polymers*, *10*(5), p.505.

[2] Norazlina, H. and Kamal, Y., 2015. Graphene modifications in polylactic acid nanocomposites: A review. *Polymer Bulletin*, *72*(4), pp.931-961.

[3] Sawyer, D.J., 2003, October. Bioprocessing–no longer a field of dreams. In *Macromolecular symposia* (Vol. 201, No. 1, pp. 271-282). Weinheim: WILEY-VCH Verlag.

[4] Gruber, P.R., Drumright, R.E. and Henton, D.E., 2000. Polylactic acid technology. *Adv. Mater*, *12*, pp.1841-1846.

[5] Dorgan, J.R., Lehermeier, H.J., Palade, L.I. and Cicero, J., 2001, October. Polylactides: properties and prospects of an environmentally benign plastic from renewable resources. In *Macromolecular symposia* (Vol. 175, No. 1, pp. 55-66). Weinheim: WILEY-VCH Verlag GmbH.

[6] Vink, E.T., Rabago, K.R., Glassner, D.A. and Gruber, P.R., 2003. Applications of life cycle assessment to NatureWorks™ polylactide (PLA) production. *Polymer Degradation and stability*, *80*(3), pp.403-419.

[7] Bajpai, P.K., Singh, I. and Madaan, J., 2013. Tribological behavior of natural fiber reinforced PLA composites. *Wear*, *297*(1-2), pp.829-840...

[8] Lee, J.T., Kim, M.W., Song, Y.S., Kang, T.J. and Youn, J.R., 2010. Mechanical properties of denim fabric reinforced poly (lactic acid). *Fibers and Polymers*, *11*(1), pp.60-66.

[9] Huda, M.S., Drzal, L.T., Mohanty, A.K. and Misra, M., 2008. Effect of fiber surface-treatments on the properties of laminated biocomposites from poly (lactic acid) (PLA) and kenaf fibers. *Composites science and technology*, *68*(2), pp.424-432.

[10] Hussain F, Hojjati M, Okamoto M, Gorga RE. Review article:polymer–matrix nanocomposites, processing, manufacturing, and application: an overview. Journal of Composite Materials 2006;40:1511–75.

[11] Laoutid F, Bonnaud L, Alexandre M, LopezCuesta JM, Dubois P. New prospects in flame retardant polymer materials: from fundamentals to nanocomposites. Materials Science and Engineering:R 2009;63:100–25.

[12] Rasal, R.M., Janorkar, A.V. and Hirt, D.E., 2010. Poly (lactic acid) modifications. *Progress in polymer science*, *35*(3), pp.338-356..

[13] Yang, B., Wang, D., Chen, F., Su, L.F., Miao, J.B., Chen, P., Qian, J.S., Xia, R. and Liu, J.W., 2019. Melting and Crystallization Behaviors of Poly (Lactic Acid) Modified with Graphene Acting as a Nucleating Agent. *Journal of Macromolecular Science, Part B*, pp.1-15.

[14] Aoyama, S.; Park, Y. T.; Ougizawa, T.; Macosko, C. W. Melt Crystallization of Poly(ethylene Terephthalate): Comparing Addition of Graphene vs. Carbon Nanotubes. Polymer 2014, 55, 2077–2085

[15] Wu, D., Cheng, Y., Feng, S., Yao, Z. and Zhang, M., 2013. Crystallization behavior of polylactide/graphene composites. *Industrial & Engineering Chemistry Research*, *52*(20), pp.6731-6739.

[16] LI, S.H., LIU, J.C., WANG, D.Y., CHEN, Y.W., WEI, H.Z., QIN, Z.M. and JIA, H.M., 2016. Research Progress on Preparation and Performance of Bio-Degradation PLA/Graphene Nanocomposites. *DEStech Transactions on Materials Science and Engineering*, (amst).

[17] KubotaM(2007) Properties of nano-structured pure Al produced by mechanical grinding and spark plasma sintering. J Alloys Compd 434:294–297

[18] Cavaliere, P., Sadeghi, B. and Shabani, A., 2019. Spark Plasma Sintering: Process Fundamentals. In *Spark Plasma Sintering of Materials* (pp. 3-20). Springer, Cham.

[19] Tanaka A, Umeda K, Takatsu S (2004) Friction and wear of diamond-containing polyimide composites in water and air. Wear 257(11):1096–1102.

[20] Schwertz M, Lemonnier S, Barraud E, Carradò A, Vallat MF, Nardin M (2014) Consolidation by spark plasma sintering of polyimide and polyetheretherketone. J Appl Polym Sci 131(18).

[21] Adesina, O.T., Sadiku, E.R., Jamiru, T., Ogunbiyi, O.F., Beneke, L.W. and Adegbola, A.T., 2019. Optimization of SPS processing parameters on the density and hardness properties of graphene reinforced polylactic acid nanocomposite. *The International Journal of Advanced Manufacturing Technology*, pp.1-12.

[22] Munir, Z.A., Anselmi-Tamburini, U. and Ohyanagi, M., 2006. The effect of electric field and pressure on the synthesis and consolidation of materials: A review of the spark plasma sintering method. *Journal of Materials Science*, *41*(3), pp.763-777.

Acknowledgement
The authors would like to extend appreciation to Tshwane University of Technology, Pretoria, South Africa for the financial support given through bursary support.

IOP Conf. Series: Materials Science and Engineering **627** (2019) 012018 doi:10.1088/1757-899X/627/1/012018

Selection of the optimal heat treatment conditions using advanced phase transformation models

Krzysztof Bzowski, Lukasz Rauch, Maciej Pietrzyk

Department of Applied Computer Science and Modelling, AGH University of Science and Technology, Krakow, Poland

kbzowski@agh.edu.pl

Abstract. The paper focuses on the problem of optimal cooling cycle selection for steel sheets after hot rolling in order to obtain a given phase composition using complex numerical model. The proposed model driven by diffusion of carbon with the moving boundary (Stefan problem) was designed and implemented on the basis of Level Set equations coupled with diffusion equation. Application of implicit method for interface evolution allows to overcome the limitations associated with the complicated representation of material microstructure, but also in a natural way reflects the impact of front curvature and influence of chemical forces on the velocity of the interface. The article discusses the details of the model as well as the applied optimization technique. Three test scenarios are analyzed and the results are discussed.

1. Introduction

Steel sheet properties mainly depends on applied heat treatment procedures after rolling. In this stage main evolution of microstructure takes places determining final properties of the product. In industrial practice design of the optimal heat treatment after hot rolling process is time consuming, expensive and very demanding task. Therefore, numerical methods are commonly used to predict obtained thermo-mechanical properties in a final product dependently on the various boundary conditions applied in a real process. Prediction of the kinetics of phase transformations and volume fractions of phases after cooling is the main task, while properly selected ratio of hard and soft phases allows to obtain a product, which meets the client's requirements. Prediction of phase transformations kinetics is possible by using phase transformation models. Historically, Johnson-Mehl-Avrami-Kolmogorov (JMAK) type equations were used for simulations of phase transformations [1]. In this approach, attention is focused on the kinetics and microstructural aspects are essentially ignored. More physical models have been recently presented based on the application of the control theory, where description of the kinetics of phase transformation in transient state is calculated by using the second order differential equation [2]. Currently, prediction of microstructure morphology evolution became possible using models based of cellular automata [3], phase field [4] and finite element [5] approaches. In this paper new model, based on Multiple Level Set coupled with Carbon Diffusion (MLSDIFF) methods, is proposed to overcome typical problems with moving boundaries.

2. Phase transformation model

Level Set Method (LSM), developed by Osher and Sethian [6], is the basis of model proposed in this paper. It is one of the most popular implicit interface representations and evolution procedure. The major advantage of LSM is natural merging and splitting of regions without grid modifications, when the interface moves – without time consuming remeshing stage. The general idea behind LSM is to move zero level set with predefined velocity υ by solving equation of motion. Velocity is given as a vector function F defined at each degree of freedom and may depend on internal or external forces. The equation of motion governing ϕ is:

$$\frac{\partial \phi}{\partial t} + F\left|\nabla \phi\right| = 0 \tag{1}$$

The velocity field for level set approach needs to be obtained from coupled solution. The main driving force behind the phase transformation during cooling of steels is carbon diffusion typically described by Second Fick's Law:

$$\frac{\partial c}{\partial t} = D(T)\frac{\partial^2 c}{\partial x^2} \tag{2}$$

where: c – carbon concentration, t – time, D - diffusion coefficient. Level set velocity function of grain boundaries was defined as constant basic normal velocity (v_n) modified by an influence of carbon concentration gradient (v_c):

$$\vec{v} = \vec{v}_{\vec{n}0} + a_1 \nabla c \tag{3}$$

where: a_1 – scaling parameters obtained in inverse analysis. The main purpose of the model is predicting phase transformation kinetics by nucleation and growth of ferrite grains. Ferrite nucleation was implemented according to the model proposed by Zhang [7]. For each nuclei separate level set solution is established which are then combined using multiple level set approach [8]. Boundary conditions are applied to zero level set as a constant carbon concentration according to GS line of phase equilibrium diagram. Bainite grains are not modelled directly, but characteristic features of this transformation, e.g. starting temperature, are determined from analytical equations. Model was implemented using deal.ii finite element library [9].

3. Optimization model

The main purpose of optimization procedure was to find optimal cooling cycle in order to achieve a given phase composition. The proposed model is used as a core of objective function and it is responsible for phase composition prediction. Biologically-inspired computational intelligence algorithms based on genetic algorithm approach were developed using *inspyred* framework [6]. The problem was approached as a typical minimization task, which goal is to obtain the lowest possible difference between values obtained from model and the reference values as follows:

$$\varepsilon = \frac{1}{n}\sum_{i}^{n}\left(f_i - f_i'\right)^2, \tag{4}$$

where: f' – phase fraction obtained from the model, f_i' – goal value of ith phase fraction. At the current stage of development, the model may predict a fractions of ferrite, martensite and bainite. Initial population was described as a series of pair: time and temperature, within the given boundary conditions. Each pair corresponds to a point in time at which the cooling rate changes. Rank-based sampling of individuals from the population, which stochastically chooses individuals from the population with probability proportional to their rank was proposed. The rank is proportional to the function of adaptation of the individual candidate. Arithmetic crossover (AX) with gaussian mutation was used to introduce more diversity to the population. The allele of each parent is weighted by the 0.1, and the allele of the complement parent is weighted by 0.9 producing new offspring with crossover rate equal 1. Gaussian mutation with probability equal 0.1 was applied for each allele. Gaussian mean was taken as 0 with standard deviation of function equal 1. In additional to this crowding replacement was used. If significant number of similar individuals have been identified, a random sample of individuals is pulled from the current population, and the closest individual to the current offspring (determined by the L2 distance) is replaced by that offspring, if the offspring is better. The following have been used as terminators for optimization: diversity less than 0.001 and no improvement of the best fitness for 10 generations.

The approach was tested for Dual Phase (DP) steels cooling, applied with six cooling rates according to the number of laminar cooling zones in real hot rolling mill. Figure 1 shows the chosen cooling scheme with characteristic turning points, constraints and boundary conditions. Times of each stage are optimization variables with ranges given in Figure 1.

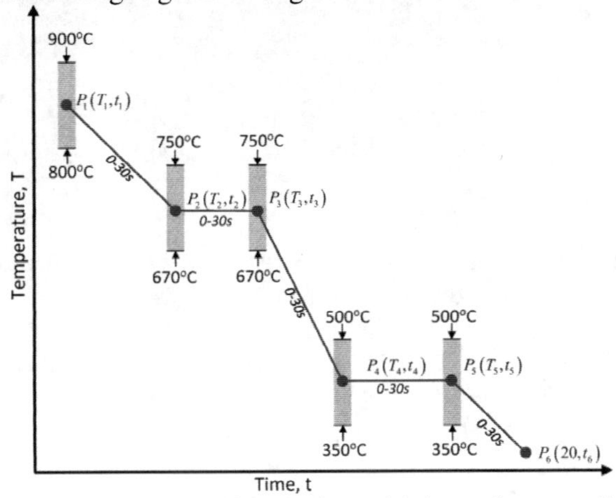

Figure 1. Laminar cooling design with boundary conditions.

4. Results and discussion

Typical DP steel containing 0.07%C was considered – three cases with different required phase compositions were analyzed and discussed i.e.: (I) 0.5 ferrite and 0.5 martensite, (II) 0.75 ferrite and 0.25 martensite, (III) 0.25 ferrite and 0.75 martensite.

Cooling schemes obtained from optimization are presented in figure 2a and a process of objective function changes is in figure 2b. Obtained results allow to design laminar cooling intensity after hot rolling process.

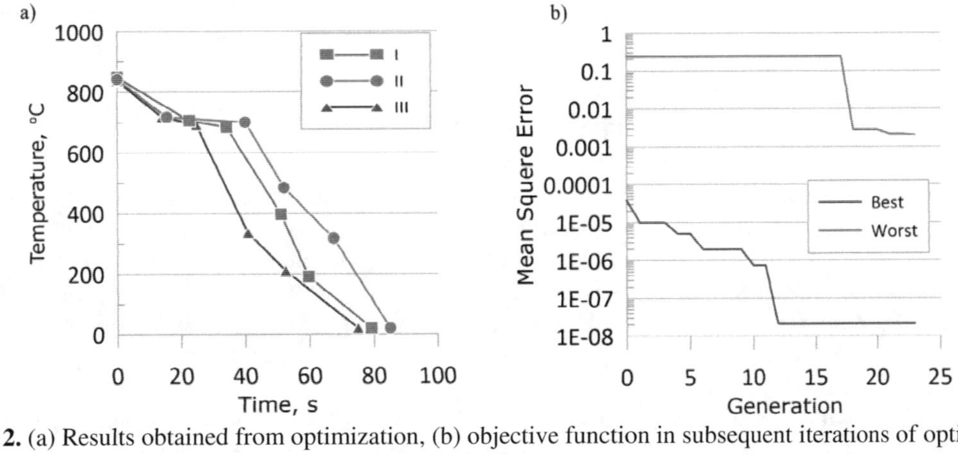

Figure 2. (a) Results obtained from optimization, (b) objective function in subsequent iterations of optimization procedure.

The representative austenitic microstructure image, 25x25μm, created using digital microstructure representation algorithms was used for the purposes of optimization process (figure 3A). Example results with carbon concentration distributation and ferrite grains morhology obtained from the final step of the optimisation procedure are shown in figures 3bc.

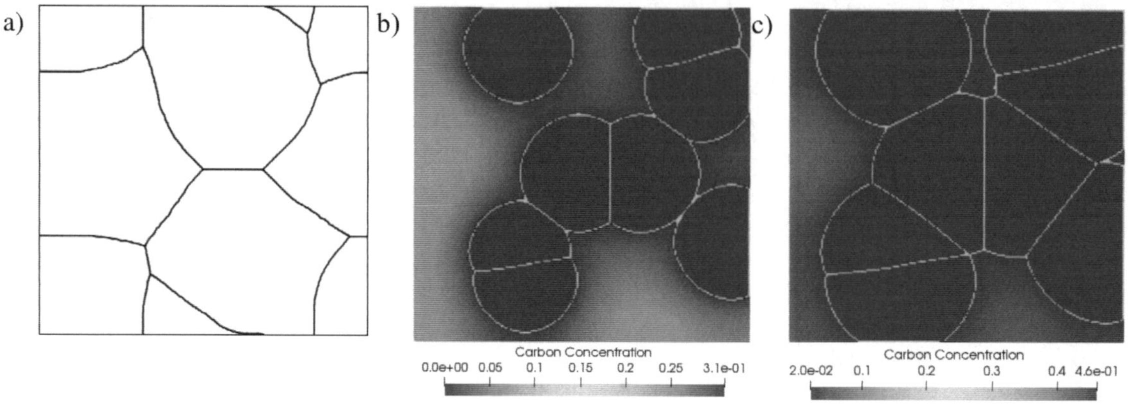

Figure 3. (a) The basic microstructure of austenite used in the optimization procedure, (b) results obtained for the test case I, (c) results obtained for the test case II.

Level set approach allows robust and efficient simulation of the moving boundary problems. Combination with carbon diffusion solution allowed to create phase transformation model capable to predict kinetics of transformation, which include morphology changes of microstructures during laminar cooling. The results obtained from the model may find practical application in multi-scale simulations taking into account the morphology of the microstructure. The proposed procedure can be used to design a laminar cooling process after hot rolling. However, due to the calculation time, it is not possible to use it online. The acceleration of calculations will be the subject of further research.

5. References

[1] Rauch Ł, Bzowski K, Perzyński K, Madej Ł, Milenin A and Pietrzyk M 2016 Strategy for the selection of the best phase transformation model for simulation of metals processing *Computer Methods in Materials Science* **16** 224-37
[2] Milenin I, Pernach M and Pietrzyk M 2015 Application of the control theory to modelling austenite-ferrite phase transformation in steels *Computer Methods in Materials Science* **15** 327-35
[3] Halder C, Madej L and Pietrzyk M 2014 Discrete micro-scale cellular automata model for modelling phase transformation during heating of dual phase steels *Archives of Civil and Mechanical Engineering* **14** 96-103
[4] Militzer M 2011 Phase field modeling of microstructure evolution in steels *Current Opinion in Solid State and Materials Science* **15** 106-15
[5] Pernach M, Bzowski K, Rauch L and Pietrzyk M 2017 Analysis of predictive capabilities of multiscale phase transformation models based on the numerical solution of heat transfer and diffusion equations *International Journal for Multiscale Computational Engineering* **15** 413-30
[6] Osher S and Sethian J A 1988 Fronts propagating with curvature-dependent speed: Algorithms based on Hamilton-Jacobi formulations *Journal of Computational Physics* **79** 12-49
[7] Zhang L, Zhang C B, Wang Y M, Wang S Q and Ye H Q 2003 A cellular automaton investigation of the transformation from austenite to ferrite during continuous cooling *Acta Materialia* **51** 5519-27
[8] Zhang X, Chen J-S and Osher S 2008 A multiple level set method for modeling grain boundary evolution of polycrystalline materials *Interaction and multiscale mechanics* **1** 191-209
[9] Alzetta G, Arndt D, Bangerth W, Boddu V, Brands B, Davydov D, Gassmöller R, Heister T, Heltai L, Kormann K, Kronbichler M, Maier M, Pelteret J-P, Turcksin B and Wells D 2018 The deal.II library, Version 9.0 *Journal of Numerical Mathematics* **26** 173-83

Acknowledgments
Financial support of the NCN project no. 2015/17/N/ST8/01024 is acknowledged.

A Semi-Analytic Solution for Plane Strain Bending Under Tension of a Strain Hardening Sheet Including Ductile Fracture Prediction

S E Alexandrov[1,a], E A Lyamina[2,b] and N Mosleni[3]

[1] Ishlinsky Institute for Problems in Mechanics RAS, 101-1 Pr. Vernadskogo, Moscow 119526, Russia
[2] Tarbiat Modares University, Jalal AleAhmad, Nasr, Tehran P.O.Box: 14115-111, Iran
[a] sergei_alexandrov@spartak.ru, [b]lyamina@inbox.ru

Abstract. A semi-analytic solution for predicting the evolution of damage in the process of plane strain bending under tension of a sheet of elastic-plastic, isotropic, incompressible, strain hardening material is provided. No restriction is imposed on the strain hardening law. The evolution of damage is described by an arbitrary uncoupled damage mechanics model. The final result is the dependence of the damage variable at the site of fracture initiation on any geometric parameter of the process in parametric form. Having the critical value of the damage variable it is possible to use this dependence for determining the value of the geometric parameter chosen at which the initiation of fracture occurs.

1. Introduction

Plane strain bending under tension at large strains is one of the classical problems in plasticity theory. Its exact analytic solution for rigid perfectly plastic material has been given in [1]. Using additional assumptions concerning the through-thickness distribution of strains analytic plane-strain solutions for more realistic models have been proposed in [2, 3]. Ignoring the transverse stress a new solution has been obtained in [4].

The present paper deals with the prediction of ductile fracture initiation in the process of plane strain bending under tension using uncoupled damage mechanics models. The solution is based on the approach proposed and developed in [5, 6]. The solution is semi-analytic.

2. Kinematics for uncoupled damage mechanics models

Let H and $2L$ be the initial thickness and width of a sheet, respectively. A cross-section of the sheet at the initial instant is illustrated in Fig. 1[a]. It is always possible to choose an Eulerian Cartesian coordinate system (x, y) such that its x- axis coincides with the axis of symmetry of the process and its y-axis with one of the sides of the initial rectangular. Then, the sides A_1B_1, C_1D_1, A_1D_1, and C_1B_1 of the initial rectangular are determined in this coordinate system by the equations

$$x = 0, \quad x = -H, \quad y = -L, \quad y = L, \tag{1}$$

respectively. Introduce a Lagrangian coordinate system (ζ, η) such that

$$\zeta = x/H \quad \text{and} \quad \eta = y/H \tag{2}$$

at the initial instant. It follows from (1) and (2) that the sides A_1B_1, C_1D_1, A_1D_1, and C_1B_1 in the Lagrangian coordinate system are determined by the equations $\zeta = 0$, $\zeta = -1$, $\eta = -L/H$ and $\eta = L/H$. Then, the equations of the curves AB, CD, AD, and CB are (Fig. 1[b])

$$\zeta = 0,\ \zeta = -1,\ \eta = -L/H \text{ and } \eta = L/H, \tag{3}$$

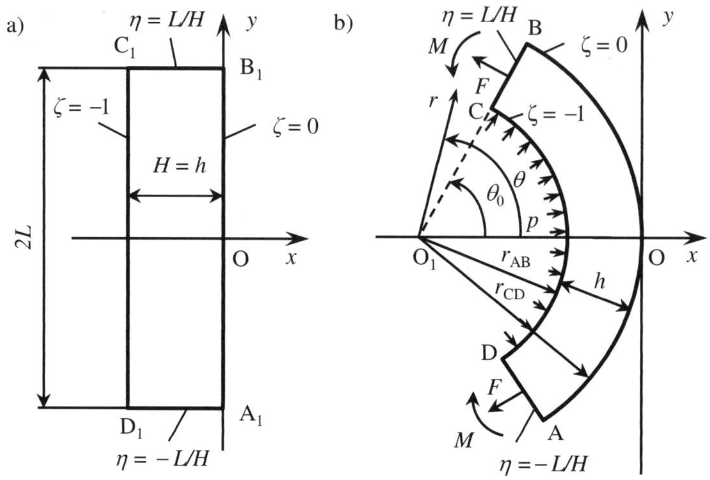

Figure 1. Bending under tension – notation.

respectively. In [5], the following mapping between the Cartesian and Lagrangian coordinate systems has been introduced for describing the process of pure bending:

$$\frac{x}{H} = \sqrt{\frac{\zeta}{a} + \frac{s}{a^2}}\cos(2a\eta) - \frac{\sqrt{s}}{a},\quad \frac{y}{H} = \sqrt{\frac{\zeta}{a} + \frac{s}{a^2}}\sin(2a\eta) \tag{4}$$

Here s is an arbitrary function of a, a is a function of the time, t, and $a = 0$ at the initial instant, $t = 0$. Moreover, the function s should satisfy the condition

$$s = 1/4 \tag{5}$$

at $t = 0$. It is evident from (4) that this mapping is symmetric relative to the x- axis. Moreover, it has been demonstrated in [5] that the mapping (4) satisfies the equation of incompressibility, transforms the sides A_1B_1 and C_1D_1 into concentric circular arcs AB and CD and the sides A_1D_1 and C_1B_1 into straight lines AD and CB orthogonal to AB and CD. The coordinate curves of the (ζ, η) coordinate system coincide with principal strain rate trajectories. Then, in the case of coaxial material models, the coordinate curves of this coordinate system coincide with principal stress trajectories. Thus the contour of the sheet is free of the shear stress throughout the process of deformation. The function $s(a)$ should be found from the solution. In the case of pure bending, this function has been found in [5] for several material models. It has been shown in [6] that the mapping (4) describes the process of bending under tension. The total principal strain components are determined from (4) as

$$2\varepsilon_\zeta = -2\varepsilon_\eta = -\ln\left[4(\zeta a + s)\right] \tag{6}$$

It is evident from (5) and (6) that $\varepsilon_\zeta = \varepsilon_\eta = 0$ at the initial instant. The further solution is significantly facilitated by introducing a moving cylindrical coordinate system (r, θ) as

$$ar = H\sqrt{\zeta a + s} \quad \text{and} \quad \theta = 2a\eta \tag{7}$$

In this coordinate system, circular arcs AB and CD are determined by the equations $r = r_{AB}$ and $r = r_{CD}$, and straight lines AD and CB by the equations $\theta = \pm\theta_0$. It follows from (7) that

$$r_{AB}/H = \sqrt{s}/a, \quad r_{CD}/H = \sqrt{s-a}/a, \quad \theta_0 = 2aL/H \tag{8}$$

3. Elastic/Plastic Solution for an Arbitrary Hardening Law

Using the mapping (4) in conjunction with the classical Eulerian theory of finite elastoplasticity [7] a semi-analytic solution for the process of plane strain bending under tension has been found in [6]. This solution is summarized in this section to introduce the nomenclature and to provide a basis for integrating damage evolution equations. The yield criterion adopted in [6] reads

$$\sqrt{3}\left|\sigma_\zeta - \sigma_\eta\right| = 2\sigma_0\Phi\left(\varepsilon_{eq}^p\right) \tag{9}$$

Here σ_ζ and σ_η are the normal stresses in the (ζ,η) system of coordinates, σ_0 is the initial yield stress in uniaxial tension, $\Phi\left(\varepsilon_{eq}^p\right)$ is an arbitrary function of the equivalent plastic strain, ε_{eq}^p, satisfying the conditions $\Phi(0)=1$ and $d\Phi\left(\varepsilon_{eq}^p\right)/d\varepsilon_{eq}^p \ge 0$ for all ε_{eq}^p. In general, the process consists of three stages. The first stage corresponds to a purely elastic solution. This stage ends when $s = s_e = \exp\left(\sqrt{3}k\right)/4$ where $k = \sigma_0/(3G)$ and G is the shear modulus of elasticity. The corresponding value of $a = a_e$ is determined from the equation

$$6kfa_e/\sqrt{s_e - a_e} = \ln^2(4s_e) - \ln^2\left[4(s_e - a_e)\right] \tag{10}$$

where $f = F/(\sigma_0 H)$ and F is the tensile force per unit length (Fig. 1b). In what follows, it is assumed that f is constant and that $a > a_e$.

A plastic region starts to develop from the surface $\zeta = 0$ at $a = a_e$. Let ζ_1 be the elastic/plastic boundary and ε_1 be the value of ε_{eq}^p at $\zeta = 0$. It has been found in [6] that

$$\begin{gathered} 4s = \exp\left\{\sqrt{3}\left[\varepsilon_1 + k\Phi(\varepsilon_1)\right]\right\}, \quad \zeta_1 = \left[\exp\left(\sqrt{3}k\right) - 4s\right]/(4a), \\ fa/\sqrt{s-a} + \ln^2\left[4(s-a)\right]/(6k) = \psi(\varepsilon_1) - \psi(0) + (k/2)\Phi^2(\varepsilon_1) \end{gathered} \tag{11}$$

Here $\psi\left(\varepsilon_{eq}^p\right)$ is an anti-derivative of $\Phi\left(\varepsilon_{eq}^p\right)$. Eliminating s in the third equation by means of the first equation leads to the equation for determining ε_1 as a function of a. Then, s and ζ_1 as functions of a can be immediately found from the first and second equation, respectively. This stage ends when another plastic region starts to develop from the surface $\zeta = -1$. The corresponding equation is

$$\exp\left(-\sqrt{3}k\right) - \exp\left\{\sqrt{3}\left[\varepsilon_1 + k\Phi(\varepsilon_1)\right]\right\} + 4a = 0 \tag{12}$$

Since ε_1 has been found as a function of a, equation (12) can be solved to determine the values of a and ε_1 at the end of the second stage. These values will be denoted by a_1 and $\varepsilon_1^{(1)}$, respectively.

There are two plastic regions, $0 \ge \zeta \ge \zeta_1$ and $\zeta_2 \ge \zeta \ge -1$, and an elastic region, $\zeta_2 \le \zeta \le \zeta_1$, at $a > a_1$. Let ε_2 be the value of ε_{eq}^p at $\zeta = -1$. It has been found in [6] that

$$\begin{gathered}2\left[\psi(\varepsilon_2)-\psi(\varepsilon_1)\right]+k\left[\Phi^2(\varepsilon_2)-\Phi^2(\varepsilon_1)\right]+2fa/\sqrt{s-a}=0,\\ 4a=\exp\left\{\sqrt{3}\left[\varepsilon_1+k\Phi(\varepsilon_1)\right]\right\}-\exp\left\{-\sqrt{3}\left[\varepsilon_2+k\Phi(\varepsilon_2)\right]\right\},\\ 4(s-a)=\exp\left\{-\sqrt{3}\left[\varepsilon_2+k\Phi(\varepsilon_2)\right]\right\},\\ 4a\zeta_2=\exp\left(-\sqrt{3}k\right)-\exp\left\{-\sqrt{3}\left[\varepsilon_2+k\Phi(\varepsilon_2)\right]\right\}-4a\end{gathered} \tag{13}$$

Using the second and third equations it is possible to eliminate a and s in the first equation. The resulting equation should be solved numerically to find ε_2 as a function of ε_1. Then, a, s and ζ_2 as functions of ε_1 can be determined from the second, third and forth equations, respectively. The equation for ζ_2 in (11) is valid and can be used to find ζ_2 as a function of ε_1.

4. Fracture prediction at the surface $\zeta = 0$

The uncoupled damage mechanics models usually involve the maximum tensile stress, σ_m, and the hydrostatic stress, σ. In particular, damage functions usually take the form

$$D=\int g(\sigma,\sigma_m)\,d\varepsilon_{eq}^p \tag{14}$$

where g is some function of its arguments. The value of D is supposed to attain its critical value, D_{cr}, at fracture. The through thickness distribution of stress has been found in [6]. However, it is not necessary to use that solution to predict the evolution of damage at $\zeta = 0$ (It is however important to know that the solution exists). Since $\zeta = 0$ is a traction free surface, it follows from (9) that

$$\sigma_m=\sigma_\eta=2\sigma_0\Phi(\varepsilon_1)/\sqrt{3} \tag{15}$$

Moreover, since the material is incompressible, $2\sigma=\sigma_\zeta+\sigma_\eta$ or using (15)

$$\sigma=\sigma_0\Phi(\varepsilon_1)/\sqrt{3} \tag{16}$$

Substituting (15) and (16) into (14) at the surface $\zeta = 0$ yields

$$D=\int_0^{\varepsilon_1} g\left[\sigma_0\Phi(z)/\sqrt{3},\,2\sigma_0\Phi(z)/\sqrt{3}\right]dz \tag{17}$$

Consider several widely used uncoupled damage mechanics models. In the case of the model proposed by Cockcroft and Latham [8] equation (17) becomes

$$D=\frac{2\sigma_0}{\sqrt{3}}\int_0^{\varepsilon_1}\Phi(z)\,dz=\frac{2\sigma_0}{\sqrt{3}}\left[\psi(\varepsilon_1)-\psi(0)\right] \tag{18}$$

In the case of the model proposed in [9], equation (17) becomes

$$D=\int_0^{\varepsilon_1}\left[1+1/\left(\sqrt{3}B\right)\right]dz=\left[1+1/\left(\sqrt{3}B\right)\right]\varepsilon_1 \tag{19}$$

Here B is a constitutive parameter. In the case of the model proposed in [10], equation (17) becomes

$$D=\int_0^{\varepsilon_1} 4/3\,dz=4\varepsilon_1/3 \tag{20}$$

In the case of the model proposed in [11], equation (17) becomes

$$D = \sqrt{3}\int_0^{\varepsilon_1}\left[\sqrt{3} - c\sigma_0\Phi(z)\right]^{-1} dz \qquad (21)$$

Here c is a constitutive parameter.

Any of the equations for D derived and (8) in conjunction with the solution given in Section 3 provide the dependence of D on r_{CD} (or any other geometric parameter). Then, the equation $D = D_{cr}$ allows the value of r_{CD} corresponding to the initiation of fracture to be calculated.

It is seen from (19) and (20) that the prediction of fracture initiation based on the models [9] and [10] is independent of the strain hardening law. The function ψ involved in (18) can be found analytically for all widely used hardening laws. The integral involved in (21) is represented in terms of hypergeometric functions for Swift's and Ludwik's hardening laws.

5. Conclusions

The general semi-analytic solution for plane strain bending under tension for elastic plastic, strain hardening material found in [6] has been adopted to predict the initiation of ductile fracture by means of uncoupled damage mechanics models. No restriction on the isotropic hardening and damage evolution laws is imposed. The final result is the dependence of the damage variable on any geometric parameter of the process in parametric form. Having a critical value of the damage variable this dependence can be used to immediately determine the value of the geometric parameter chosen at which the initiation of ductile fracture occurs.

6. Acknowledgments

This research was supported by the grants RFBR-17-58-560005 (Russia) and 96004204 (Iran National Science Foundation).

7. References

[1] Hill R 1950 *The Mathematical Theory of Plasticity* (Oxford : The Clarendon Press).
[2] Yoshida M, Yoshida F, Konishi H and Fukumoto K 2005 *Int. J. Mech. Sci.* **47** 1885
[3] Parsa M H and Ahkami S N A 2008 *Int. J. Mater. Form. Suppl.* **1** 173
[4] Guler M A, Ozer F, Yenice M and Kaya M 2010 *Steel Res. Int.* **81** 801
[5] Alexandrov S, Kim J-H, Chung K and Kang T-J 2006 *J. Strain Anal. Engng. Des.* **41** 397
[6] Alexandrov S, Manabe K and Furushima T A 2011 *Arch. Appl. Mech.* **81** 1935
[7] Xiao H, Bruhns O T and Meyers A 2006 *Acta Mech.* **182** 31
[8] Cockcroft M G and Latham D J 1968 *J. Inst. Metals* **96** 33
[9] Oyane 1972M *Bull. JSME* **15** 1507
[10] Brozzo P, Deluca B and Rendina R 1972 *Proc. Seventh Biennial Congress of International Deep Drawing Research Group* (Amsterdam, Netherlands) p 9
[11] Norris D M, Reaugh J E, Moran B and Quiñones D F 1978 *Trans. ASME J. Eng. Mater. Technol.* **100** 279

IOP Conf. Series: Materials Science and Engineering **627** (2019) 012020 doi:10.1088/1757-899X/627/1/012020

Push-Over Analysis of RC Frame with Corroded Rebar

Piero Colajanni[1], Antonino Recupero[2] and Nino Spinella[3]*

[1]Department of Engineering, University of Palermo, Viale delle Scienze, Palermo (Italy)

[2]Department of Engineering, University of Messina, C.da di Dio 1, Messina (Italy)

[3]Department of Engineering, University of Messina, C.da di Dio 1, Messina (Italy)

* nino.spinella@unime.it

Abstract. As known, the Italian building heritage largely consists of reinforced concrete frames designed before the '80s, which are, in many cases, built in the absence of specific anti-seismic criteria. Moreover, many of them, today, are characterized by bad structural conditions. Moreover, the problem of the structural conditions of the existing buildings, and their residual strength capacity, is often linked to the deterioration induced by the corrosive phenomena, which end up having a big impact on steel rebar mechanical properties. In this work, in order to investigate the influence of corrosion-damage on seismic response of existing reinforced concrete structures, a study has been carried out by analysing the non-linear behaviour of a reinforced concrete frame. The strength deterioration and reduction of the cross-section of steel rebar have been investigated and taken into account in the numerical analysis. This work shows the way in which the corrosion levels affected the push-over response, and the numerical results have been deeply analysed.

1. Introduction

The corrosion of steel reinforcement heavily affects the performances of Prestressed (PC) [1,2] and Reinforced Concrete (RC) [3] structures. It causes loss of load-carrying capacity of tendon and steel rebar, modifying both structural behaviour and capacity of elements [4,5].
In literature, there are several research works about the detrimental effects of corrosion in structures, because the problem is largely widespread [6] and the retrofitting is expensive [7].
Furthermore, it is important to develop adequate models to represent material degradation into seismic behaviour simulation of RC structures [8–11].
In this work, the seismic behaviour of a corroded framed RC structure is analysed by means of push-over analyses, which allow understanding the development of the global behaviour of the structure [12]. A degree of corrosion penetration has been simulated, by means of the reduction of bars diameters and concrete cover cracking and spalling.
At the begin, a database of experimental results on corroded steel rebar, available in literature, has been collected, and experimental laws have been calibrated to reproduce the corrosion-damage on mechanical proprieties of steel rebar.
Finally, the suggested experimental laws have been used to investigate the corrosion changes the curvature – bending moment relationship of RC cross-section, and the push-over response of a 2D RC frame.

2. Detrimental effects of corrosion on steel rebar mechanical properties

The damage-corrosion effects on the mechanical properties of materials involved were experimentally observed in different research works [5,13–19].

2.1 Concrete

The cross-section of rebar is reduced by corrosion. Moreover, the rust forms all around the cross-section of reinforcement, then a greater volume than the original thickness of rebar is occupied. Therefore, a round tensile stresses in the surrounding concrete is generated, then cracking of the cover and reducing of the compressive strength of concrete is observed.

This undesirable effect of corrosion has to take into account, thus the cylindrical compressive strength of corrosion-damaged concrete ($f_{cm,c}$) is calculated as follows [20]:

$$f_{cm,c} = f_{cm}/(1 + K\,\varepsilon_1/\varepsilon_{co}) \tag{1}$$

where $K = 0.1$ is the coefficient related to rebar roughness and diameter, ε_{c0} is the strain at the peak concrete stress f_{cm}; ε_l is the average tensile strain in cracked concrete at orthogonal direction of the applied compression.

The tensile strain can be evaluated as follows:

$$\varepsilon_1 = (b_{w,c} - b_w)/b_w \tag{2}$$

being b_w the width of the cross-section without corrosion; $b_{w,c}$ the increased width when corrosion cracks occur. In order, to give an estimation of the width increase, it can be calculated as $n_{bar}\, w_{cr}$, with n_{bar} = number of rebar in the compressed layer and w_{cr} = total crack width for a given corrosion level. The latter that can be evaluated as following [21]:

$$w_{cr} = \sum_i u_{i,corr} = 2\pi(u_{rs} - 1)Q_c \tag{3}$$

with u_{rs} = ratio volumetric expansion of the oxides (= 2), Q_c = corrosion depth and $u_{i,corr}$ = opening of single crack.

2.2 Steel Rebar

The corrosion-damage involves the steel rebar at the cross-section level by causing a reduction of steel thickness, which is interested by penetration of corrosion products. As consequence, a decrease of mechanical properties of steel rebar may be detected.

The spreaded corrosion effects may be taken into account by estimating the residual cross-sectional area of corroded rebar, i.e. reducing the bond strength [22]. However, the oxidation products of pitting corrosion are less wide, thus no signs of longitudinal cracking might become visible prior to significant cross-section loss [13].

In order, to investigate the corrosion effects on mechanical properties of steel reinforcement, Cairns et al. [13] carried out a series of tests on deteriorated steel rebar. The corrosion damage was simulated by eliminating part of rebar cross-section.

On the basis of the experimental results, the degradation of mechanical properties due to the corrosion effects, was reproduced by suggesting a linear equation:

$$p_c = p(1 - \alpha_p Q_c) \tag{4}$$

where the generic property (p) can, alternatively, be: the yield strength ($f_{y,c}$), the ultimate tensile strength ($f_{u,c}$) or strain ($\varepsilon_{u,c}$) corresponding to ultimate strength after corrosion; f_y, f_u and ε_u are yield strength, ultimate strength and strain of the non-corroded rebar, respectively; and $\alpha_y = 0.012$, $\alpha_u = 0.011$ and α_l = 0.030 are empirical coefficients obtained by a statistical analysis.

IOP Conf. Series: Materials Science and Engineering **627** (2019) 012020 doi:10.1088/1757-899X/627/1/012020

The empirical coefficients suggested by Cairns et al. [13] were calibrated on own experimental results. With the aim to improve the performances of the linear detrimental models, a database of tests, carried out on corroded steel rebar subjected to tensile stress, has been collected.
Three test campaigns, for a total of 164 specimens, have been chosen [5,6,14]. In each test campaign, different degrees of corrosion were achieved. They were obtained by varying the application time of average current density. Then, the longer the current was applied the greater the corrosion effects were in rebar.
The specimens had variable diameter, ranging from 6 to 20 mm. The mechanical properties investigated have been: the yield strength ($f_{y,c}$), the ultimate strength ($f_{u,c}$), and the ultimate strain ($\varepsilon_{u,c}$) of corroded steel rebar.
By analysing the experimental test results, it was observed that the steel rebar mechanical properties were degraded with the increase of the corrosion degree. Then, corroded steel rebar was less ductile and strength than the non-corroded rebar, as expected.
Therefore, the experimental results have been used to retuned the empirical coefficients, by taking also in account the role of the rebar diameter (ϕ):

$$p_c = p(1 - \alpha_p \phi Q_c) \tag{5}$$

The values of the coefficients, obtained by the fitting procedure, have been: $\alpha_y = 0.00055$; $\alpha_u = 0.00057$; and $\alpha_l = 0.00214$. The calculated values are different from the analogues suggested by Cairns et al. [13]. It is due to the fact that the decay law has been modified introducing also the diameter of the rebar.
In Figure 1 the experimental versus numerical values, for each mechanical property ratio ($f_{y,c}/f_y$, $f_{u,c}/f_u$, and $\varepsilon_{u,c}/\varepsilon_u$), are shown. The points have grouped all around the optimal line ($exp = num$) for both $f_{y,c}/f_y$ and $f_{u,c}/f_u$, while are slightly dispersed for $\varepsilon_{u,c}/\varepsilon_u$. It is because the ultimate strain of corroded steel rebar is influenced by several parameters, then a different equation form for the regression curve should be used. Moreover, most of the values for the ratios $f_{y,c}/f_y$ and $f_{u,c}/f_u$ are within the cone ±20% and the statistical coefficients, Mean and Coefficient of Variation (CoV), state that the assumed linear law fits with enough precision the experimental data.

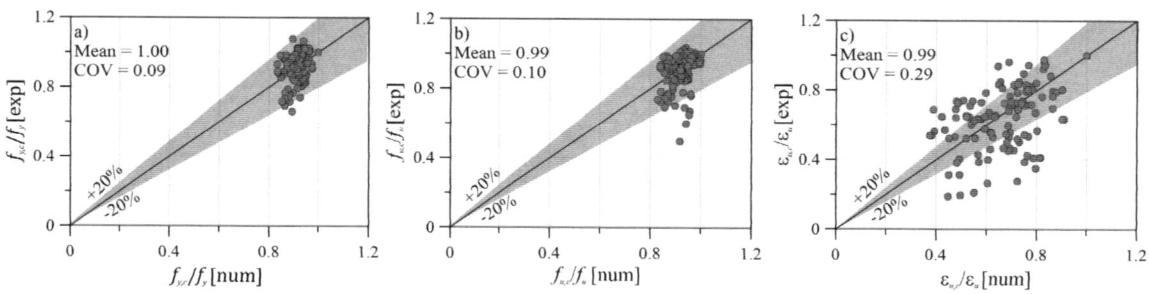

Figure 1. Experimental versus numerical value for: a) $f_{y,c}/f_y$; b) $f_{u,c}/f_u$; and c) $\varepsilon_{u,c}/\varepsilon_u$.

3. Curvature – bending moment relationship of corroded RC cross-section

Before to perform the push-over analysis of a corroded RC frame, the influence of corrosion on the cross-section properties of beams and columns has been investigated. It is known that the global non-linear static behaviour of RC structures is strictly related to the cross-section properties. Moreover, in this study, the push-over analysis has been performed, as it is explained in the following section, by using the classical theoretical model of concentrated plastic hinge.
The plastic hinge is modelled by the curvature - bending moment (χ-M) relationship and normal force-bending moment (N-M) interaction diagram (this last only for columns).
Because, the RC frame analysed has been part of a real corroded RC structure, built in 70's in the Messina (Italy) seismic area, all cross-sections have been modelled, assuming a corrosion degree that

decrease the material mechanical properties, on the basis of the predictive model described in the previous section [Eq. (5)].
Two RC cross-sections have here been reported. The geometry of beam is 300×700 mm, with a flexural reinforcement of 3ϕ20 + 2ϕ14 in compression (A_{top}) and 2ϕ18 + 2ϕ14 in tension (A_{bot}). The geometry of column is 300×800 mm and the longitudinal reinforcement is 3ϕ22 both on top and bottom of cross-section.
The corrosion degree has been assumed equal to 15%, then the suggested linear model allows to estimate the mechanical properties of corrosion-damaged materials. They are listed in Table 1, together with the non-corroded ones and the percentage decay (Δ) due to the corrosion effects.

Table 1. Mechanical properties of materials without and with corrosion.

Corrosion degree [%]	f_{cm} [MPa]	f_y [MPa]	f_u [MPa]	ε_u [mm/m]
0	10.7	417.6	590.9	0.219
15	8.9	360.0	512.4	0.107
Δ	17%	14%	13%	51%

The estimation of mechanical properties is preparatory to calculate the curvature – bending moment of cross-section. The plastic hinge of beams and columns has been characterized by using a software, namely RC-ABC [23], based on a sectional-fibre model [24].

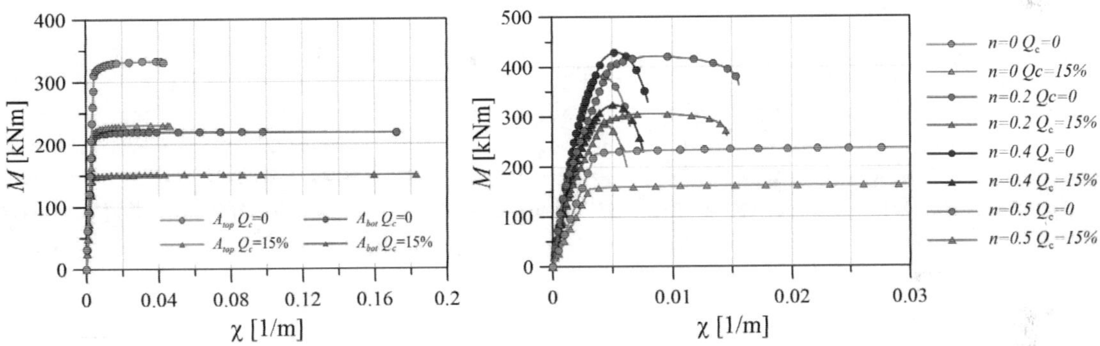

Figure 2. a) Curvature – Bending Moment relationships for RC beam. B) Curvature – Bending Moment relationships for RC column.

In Figure 2, the curvature - bending moment curves for beam and column are shown. For RC beam, two cases have been considered: a) with top reinforcement in tension; and b) with bottom reinforcement in tension. For RC column, four values of non-dimensional axial load (n) have been considered: 0, 0.2, 0.4, and 0.5.
The corrosion degree influences both the maximum bending moment and curvature. These values are decreased and, in both cases, the ductility also.
In Figure 3, the *N-M* interaction domain of column is plotted. It is evident as the corrosion drastically decreases the combined loads bearing capacity of the column (the shadow area in Figure 3).
A more sophisticate model than the used linear law [Eq. (5)] may allow to obtain more refined numerical results. However, the improvements in terms of χ-M curves and *N-M* interaction domains would be of little significance if compared to the complication of a more sophisticate model.
Furthermore, the estimation of steel rebar ultimate strain (ε_u) influences the non-linear behaviour of RC cross-section, by getting a brittle or ductile failure mode. To investigate this topic, further numerical analysis should be carried out.

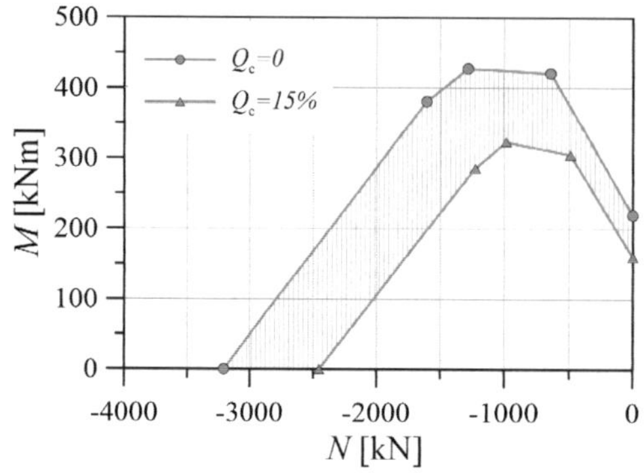

Figure 3. Axial load – Bending Moment interaction domain for RC column.

4. Push-Over Response of Corroded RC Frame

A push-over analysis has been carried out, by using the SAP2000 software, on an RC plane frame. It has been part of an existent structure, that was built in 70's in the Messina strait (Italy) area.
The general structural conditions of building are good, except for a diffused corrosion of beams and columns. In particular, as regards foundation and structural elements exposed to the atmospheric events are interested by corrosion-damage. Then, a lateral RC frame has been choose for the analysis.
The non-linear static analyses has been performed by implementing the numerical plastic rotation – bending moment relationships for each beams and columns of RC frame.
Two different scenarios were assumed: the non-corroded plane frame, and the plane frame interested by a 15% corrosion degree. The empirical model suggested before to predict the detrimental effect of corrosion on material mechanical properties has allowed to model all plastic hinges in the RC frame.
The Figure 4 shows the push-over analysis results of non-corroded and corroded frame. As can be observed the distribution of plastic hinge is quite the same. In the non-corroded frame, a lower number of plastic hinges than corroded frame has been observed. Furthermore, the average exploitation level of plastic hinges was higher, as expected.

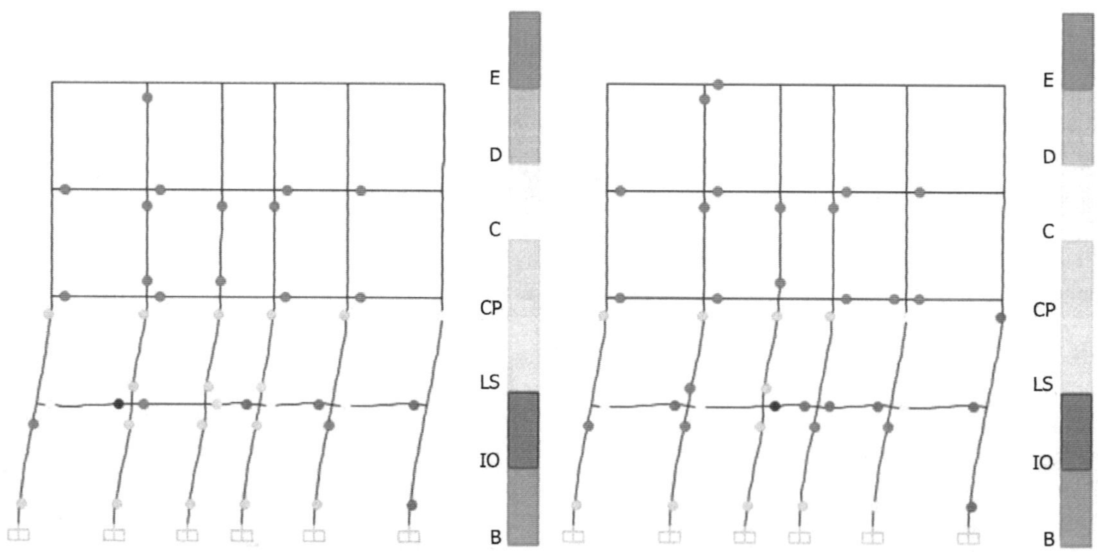

Figure 4. Last step of push-over analyses of: a) non-corroded; and b) corroded RC frame.

In Figure 5, the push-over curves for plain and corroded frame, respectively, are plotted.
These curves highlight the corrosion effects influence on the base shear-behaviour of RC frame. The corroded RC frame has been less ductile than not-corroded one, by tracing the less strength and ductility of materials involved. The area under the δ-V curve is small for the corroded RC frame, as consequence of the reduced dissipation capacity.

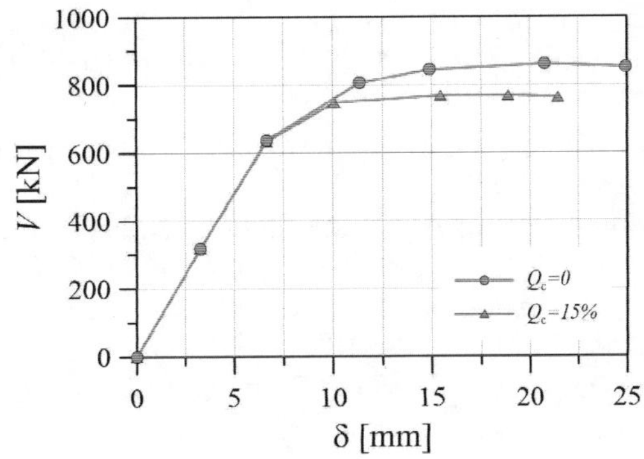

Figure 5. Displacement – Total base shear curve.

6. Conclusions

In this paper, the corrosion effects in the push-over response of RC frames have been analysed.
First, a database of test on corroded rebar has been collected, thus the corrosion-damage on mechanical properties of steel rebar has been investigated.
By examining these data, a simple equation for the decay prediction of material mechanical properties, as function of the degree of corrosion and the rebar diameter, has been provided. The consistency of this linear relationship has been discussed and validated, highlighting the need of further investigations with the aim of better reproducing the ductile behaviour of corroded RC cross-section.
Some numerical analysis have been carried out to evaluate the curvature - bending moment relationships of RC cross-sections, characterized by corrosion phenomena. They show that the corrosion drastically degreases the ductility and ultimate strength of RC cross-section.
Finally, the non-linear static behaviour of an existing RC frame, which is part of a real structure characterized by corrosion, has been evaluated.
The distribution of the plastic hinges remain quite the same as the non-corroded case. By contrast, the corrosion effects on the global behaviour of the RC frame lead to less ductility and strength, confirming the need to further investigate this issue with the aim to propose more refined models.

7. References

[1] Colajanni P, Recupero A, Ricciardi G and Spinella N 2016 Failure by corrosion in PC bridges: A case history of a viaduct in Italy *Int. J. Struct. Integr.* **7** 181–93

[2] Recupero A and Spinella N 2019 Preliminary results of flexural tests on corroded prestressed concrete beams *Proceedings of the fib Symposium 2019: Concrete - Innovations in Materials, Design and Structures* (Krakow, Poland) pp 1323–30

[3] Cesetti A, Mancini G, Tondolo F, Recupero A and Spinella N 2016 Physical model for structural evaluation of R.C. beams in presence of corrosion *Proceedings of the 4th International Conference on Concrete Repair, Rehabilitation and Retrofitting, ICCRRR 2015* ed D Frank, H-D Beushausen, Mark G Alexander and Pilate Moyo (Leipzig, Germany) pp 107–14

[4] Mancini G, Tondolo F, Iuliano L and Minetola P 2014 Local reinforcing bar damage in r.c. members due to accelerated corrosion and loading *Constr. Build. Mater.* **69** 116–23

[5] Fernandez I, Bairán J M and Marí A R 2016 Mechanical model to evaluate steel reinforcement corrosion effects on σ–ε and fatigue curves. Experimental calibration and validation *Eng. Struct.*

118 320–33

[6] Almusallam A A, Al-Gahtani A S, Aziz A R, Dakhil F H and Rasheeduzzafar 1996 Effect of Reinforcement Corrosion on Flexural Behavior of Concrete Slabs *J. Mater. Civ. Eng.* **8** 123–7

[7] Spinella N, Colajanni P, Recupero A and Tondolo F 2019 Ultimate Shear of RC Beams with Corroded Stirrups and Strengthened with FRP *Build. 2019, Vol. 9, Page 34* **9** 34

[8] Colajanni P, Recupero A and Spinella N 2015 Shear strength degradation due to flexural ductility demand in circular RC columns *Bull. Earthq. Eng.* **13** 1795–807

[9] Colajanni P, Recupero A and Spinella N 2014 Design procedure for prestressed concrete beams *Comput. Concr.* **13** 235–53

[10] Colajanni P, La Mendola L, Monaco A and Spinella N 2016 *Cyclic behavior of composite truss beam-to-RC column joints in MRFS* vol 711

[11] Colajanni P, La Mendola L, Recupero A and Spinella N 2017 Stress field model for strengthening of shear-flexure critical RC beams *J. Compos. Constr.* **21**

[12] Spinella N 2019 Push-over analysis of a rubble full-scale masonry wall reinforced with stainless steel ribbons *Bull. Earthq. Eng.* **17** 497–518

[13] Cairns J, Plizzari G A, Du Y, Law D W and Franzoni C 2005 Mechanical Properties of Corrosion-Damaged Reinforcement *ACI Mater. J.* **102** 256–64

[14] Cobo A, Moreno E, Cánovas M F and Cánovas M F 2011 Mechanical properties variation of B500SD high ductility reinforcement regarding its corrosion degree *Mater. Construcción* **61** 517–32

[15] Apostolopoulos C A, Demis S and Papadakis V G 2013 Chloride-induced corrosion of steel reinforcement – Mechanical performance and pit depth analysis *Constr. Build. Mater.* **38** 139–46

[16] Imperatore S, Rinaldi Z and Drago C 2017 Degradation relationships for the mechanical properties of corroded steel rebars *Constr. Build. Mater.* **148** 219–30

[17] Zhang W, Song X, Gu X and Li S 2012 Tensile and fatigue behavior of corroded rebars *Constr. Build. Mater.* **34** 409–17

[18] Recupero A, Spinella N and Tondolo F 2018 Failure analysis of corroded RC beams subjected to shear-flexural actions *Eng. Fail. Anal.* **93**

[19] Recupero A, Spinella N and Tondolo F 2018 A model for the analysis of ultimate capacity of RC and PC corroded beams *Adv. Civ. Eng.*

[20] Coronelli D and Gambarova P 2004 Structural Assessment of Corroded Reinforced Concrete Beams: Modeling Guidelines *J. Struct. Eng.* **130** 1214–24

[21] Molina F J, Alonso C and Andrade C 1993 Cover cracking as a function of rebar corrosion: Part 2—Numerical model *Mater. Struct.* **26** 532–48

[22] Giordano L, Mancini G and Tondolo F 2009 Numerical Interpretation of Bond Between Steel and Concrete in Presence of Corrosion and Cyclic Action *Key Eng. Mater.* **417–418** 349–52

[23] Lignola G P, Giamundo V, Prota A and Manfredi G 2013 FRP wrapping of RC members under combined axial load and bending *11th International Symposium on Fiber Reinforced Polymer Reinforcement for Concrete Structures* (Guimarães, Portugal) p 06_639-1-06_639-10

[24] Spinella N 2013 N-M-χ interaction for arbitrary cross section under biaxial bending and axial load *Pollack Period.* **8** 87–100

IOP Conf. Series: Materials Science and Engineering **627** (2019) 012021 doi:10.1088/1757-899X/627/1/012021

Anti-sliding Risk Model of Wet Surface for Asphalt Pavement and Test Verification

Ke Zhong, Mingzhi Sun*, Zhaoxu Liu, Xiaochen Ma, Fei Chen, Rong Ma

Research Institute of Highway Ministry of Transport, Beijing, 100088, China

*Corresponding author's e-mail: mz.sun@rioh.cn

Abstract. A layer of water film will form on the road surface in rainy weather. Due to the lubrication effect of water film, the anti-skid ability of driving vehicles decreases, which will lead to road traffic accidents. The contact between water film, tread rubber and pavement surface texture is the most direct and concrete when a vehicle travels on wet sliding road surface. Their respective characteristics and coupling effects provide the basis and essential source of tire-road friction. Based on this, in this paper a tire-road anti-skid safety risk model under rainy weather with friction coefficient as evaluation criterion was established by analyzing the fractal characteristics of asphalt pavement surface texture and the viscoelastic properties of rubber tire. Then the correctness of the model was verified by DF test. The test results show that the trend of friction curve calculated by the model is similar to that measured by DF test, which verifies that the model can be used to calculate and characterize the anti-sliding performance of road surface in rainy weather.

1. Introduction

With the rapid development of the highway industry, transportation demand has increased significantly, and traffic volume has increased year by year. At the same time, the speed of modern cars continues to increase, resulting in frequent traffic accidents. Traffic safety is closely related to the anti-skid performance of asphalt pavement. Defects of anti-skid performance of asphalt pavement can lead to slippage of vehicles on the pavement or traffic accidents caused by too long emergency braking distance. How to accurately evaluate the anti-skid performance of pavement is particularly critical. In order to accurately evaluate the anti-sliding performance of the road, the researchers used theoretical calculations, experimental research and finite element modeling to evaluate the anti-sliding performance evaluation methods and indicators, anti-sliding mechanism, anti-sliding performance factors and anti-sliding performance decay laws. Establish a systematic, objective and scientific evaluation system to ensure the anti-sliding performance of asphalt pavement and maximize the anti-sliding performance of vehicles on asphalt pavement. Muarat Ergun [1] comparatively analyzed the measured data and the friction-related characterization values obtained from the surface texture data of the pavement, quantitatively described the relationship between the friction coefficient of the road surface and the macro-texture and micro-texture, and established the quantitative relationship suitable for the common commonly used graded mixture. Marsac [2] uses the tire-pavement finite element model to study the effect of road texture on friction coefficient and gives the curve of friction coefficient varying with speed. Persson [3-4] put forward the theory of friction contact between rubber and self-affine fractal hard cushion. According to this theory, the rough property of hard cushion, the viscoelasticity and the interaction of rubber are the friction source in this particular case. Based on this theory, the theoretical model of friction coefficient has been put forward, which shows that pavement texture has fractal characteristics. Fractal dimension can be used to describe the complexity and similarity of pavement texture [5-6]. It provides a new idea for describing the relationship between pavement texture characteristics and anti-skid performance. When there is water film on the pavement, Ji Tianjian studied the thickness of pavement water film under rainfall conditions by simulation test method. The relationship between pavement water film thickness and rainfall intensity, slope length, slope and rough surface roughness was studied through experimental data [7]. Cong Ling et al. combined with fluid dynamics theory and finite element method to propose a method to determine the braking distance when considering the friction characteristics of the road surface [8]. Fwa et al. simulated the tire friction effect on the wet slip surface by three-dimensional finite element model, and analyzed the influence of water film thickness on the friction coefficient by numerical method [9-10].

At present, the models describing road friction coefficients are regression models or finite element models based on measured data for analysis. There are great differences in measuring conditions, regression methods and model building conditions, which limit their generality. Tread rubber and pavement surface texture are the most direct and specific contact parts in the tire-road friction system. Their respective characteristics and their coupling are the material basis and essential source of the friction system. In this paper, by exploring the characteristics of tire viscoelasticity and road surface texture morphology and their interaction mechanism, a safety risk model for wet slip asphalt surface is established based on friction coefficient. Finally, the friction coefficient curve of asphalt pavement samples was obtained by DF test to verify the correctness of the anti-sliding safety risk model of wet asphalt surface.

2. Calculation of power spectrum on road surface

The surface of asphalt pavement has fractal characteristics. The size of fractal dimension can reflect the complexity and richness of surface texture comprehensively. Moreover, the asphalt surface texture section elevation data can be regarded as a one-dimensional signal, so the texture characteristics can be studied by means of signal processing.The power spectrum can reflect the energy distribution of the signal in the frequency domain, and the statistical distribution of the surface elevation value is characterized by the power spectrum of the texture. The power spectrum calculation theory of fractal surface is used to represent the texture characteristics of asphalt pavement. The approximate calculation of power spectrum of asphalt pavement is shown in formula (1):

$$C(q) \approx \frac{H}{2\pi} \cdot \left(\frac{h_0}{q_L}\right)^2 \cdot \left(\frac{q}{q_L}\right)^{-2(H+1)} \tag{1}$$

h_0 is the root mean square texture elevation of Asphalt Pavement, q is the wave vector of Asphalt Pavement, ql is the upper limit wave vector of Asphalt Pavement, qL is the lower limit wave vector of Asphalt Pavement, H is House index.

3. Tire-road anti-sliding risk model

3.1. Model assumption

The surface of the road will be covered by water film during rainfall.It is found that the hydrodynamic effect on the tire-road coupling can be neglected when there is a thin water film on the road surface. When the tire slides on the road surface, there is enough time to drain the water between the tire and the road. Only part of the water will be left in the crack of the road surface by rubber seal, forming a closed "pool" phenomenon. The rubber particles of the tire can fully contact with the bulge of the asphalt pavement, but can not fully contact with the bottom of the groove. When the tread rubber slides on the asphalt pavement, λ defined as the wavelength, A(λ) is the true contact area, A(L) is the ideal contact area for sufficient contact and L is the radial length of the ideal contact area. In order to characterize the case where the roughness of the tread rubber and asphalt pavement is superior, the proportional coefficient P(q)=P(ζ)=A(λ)/A(L) is introduced to characterize the ratio of the true contact area at the wavelength ζ scale and the ideal contact area at the full contact. The proportional coefficient P(q) is represented by the following formula.

$$P(q) = \frac{2}{\pi}\int_0^\infty dx \frac{\sin x}{x} \exp[-x^2 G(q)] \tag{2}$$

The function G(q) is represented by the Eq.(3).

$$G(q) \approx \frac{H}{16\pi} \cdot h_0^{\ 2} \cdot q_L^{\ 2H} \cdot \int_{q_L}^{q} d\zeta \zeta^{-2H+1} \times \int_0^{2\pi} d\varphi \left| \mathrm{Im} \frac{E(q_L \zeta v \cos\varphi)}{(1-\gamma^2)\sigma_0} \right|^2 \tag{3}$$

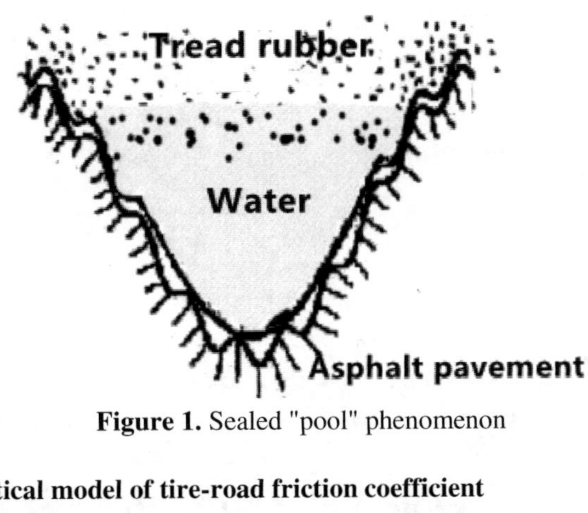

Figure 1. Sealed "pool" phenomenon

3.2. Construction of theoretical model of tire-road friction coefficient

When tread rubber and asphalt pavement are fully in contact with sliding, A0 is the ideal contact area, σf is the frictional stress between the tire and the road surface, v is the sliding speed, γ is the tread rubber Poisson's ratio, t0 is the sliding time, u is the displacement, and σ is the stress distribution function. The energy consumed in the sliding time t0 is represented by the Eq.(4).

$$\Delta E = \sigma_f A_0 v t_0 = \int dx^2 dt\, u\, \sigma \tag{4}$$

The displacement function uz of the elevation direction is denoted by h, and the σf expression can be derived from the Eq.(4).

$$\sigma_f = \frac{(2\pi)^2}{vA_0} \int dq^2 (iw)[M_{zz}(-\bar{q},-w)]^{-1} \times < h(\bar{q}) h(-\bar{q}) > \tag{5}$$

IOP Conf. Series: Materials Science and Engineering **627** (2019) 012021 doi:10.1088/1757-899X/627/1/012021

$M_{zz}(\bar{q},w)\approx-\frac{2(1-\gamma)}{Eq}$ is stiffness matrix and $<h(\bar{q})h(-\bar{q})>=\frac{A_0}{(2\pi)^2}C(q)$ is correlation coefficient.

The expressions of stiffness matrix and correlation coefficient are introduced into equation (5) to obtain sufficient expressions for contact sliding:

$$\sigma_f=\frac{1}{2}\int dq^2\cdot q^2\cos\varphi\cdot C(q)\,\mathrm{Im}\frac{E(qv\cos\varphi)}{1-\gamma^2} \tag{6}$$

The maximum wave vector of the asphalt pavement is ql, and the minimum value is qL. The surface topography of the pavement has a sinusoidal function curve on any section. Rotating 2π over the horizontal plane is the corresponding surface, so the angle between the wave vector and the velocity is [0,2π]. By the definition of the friction coefficient, the proportional coefficient B is introduced at the same time to obtain the friction coefficient expression under normal conditions:

$$\mu=\int_{q_L}^{q_l}dq\cdot q^3C(q)P(q)\times\int_0^{2\pi}d\varphi\cos\varphi\,\mathrm{Im}\frac{E(qv\cos\varphi)}{(1-\gamma^2)\sigma_0} \tag{7}$$

Substituting equations (1), (2) and (3) into equation (7) to obtain the anti-sliding calculation model for wet-slip asphalt surface.

$$\mu\approx\frac{1}{4\pi}(q_Lh_0)^2H\int_1^{q_l/q_L}d\zeta\zeta^{-2H+1}P(q)$$
$$\times\int_0^{2\pi}d\varphi\cos\varphi\,\mathrm{Im}\frac{E(\zeta q_L v\cos\varphi)}{(1-\gamma^2)\sigma_0} \tag{8}$$

Where $E(\zeta q_L v\cos\varphi)$ is as expressed in Eq.(8):

$$E(q_L\zeta v\cos\varphi)=\frac{E_1(1-i\tau\cdot qv\cos\varphi)}{a_1+1-i\tau\cdot qv\cos\varphi} \tag{9}$$

4. DF test verification feature model

In order to verify the rationality of the anti-slip risk model, the SMA-13 rutting plate was specially prepared. The most advanced self-focusing three-dimensional surface measuring instrument was used to extract the three-dimensional characteristic parameters of the sample surface. The test piece texture information is shown in Figure 2. Bring the parameters obtained from the test into the equation (8) for simulation calculation, and obtain the theoretical calculation results. The theoretical calculation results and the results of the DF instrument are shown in Figure 3.

Figure 2. Specimen texture information

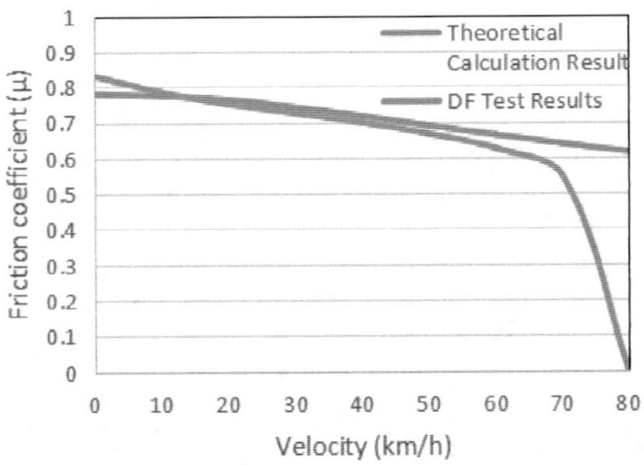

Figure 3. Comparison between theoretical calculation results and DF measured results

As can be seen from Figure 3, the final section of the measured curve of the DF meter covers the transition process between the surface of the sample and the rubber slider of the DF instrument, which is separated by contact, starts to contact, and gradually contacts until the contact is fully sufficient. The speed starts to decrease from 80km/h, and the friction coefficient grows rapidly after creeping, which is caused by test methods and conditions. After eliminating the inevitable abnormal factors at the end of the measured curve, the theoretical calculation curve and the measured curve are basically the same in the range of 0km/h to 80km/h.

5. Conclusions

This paper deduces a risk model of anti-skid on wet road surface, which can calculate the friction coefficient of vehicles when they slip on wet road surface. Then, the results of model calculation are analyzed and verified by DF real test pieces. The verification results show that the theoretical calculation results of the anti-sliding risk model are basically consistent with the measured results of the DF test. The model provides a scientific and reliable method for anti-sliding in the process of running on the wet slide surface, which has important theoretical significance and application value.

6. References

[1] M Ergum, S Iyinam (2005). Prediction of road surface friction coeffcient using only macro and micro texture measurements. Journal of transportation engineering, 131(4): 311-319.

[2] MT Do, Y Delanne (2004). Prediction of tire/wet road friction from road surface micro-texture and tire rubber properties.5th Symposium on Pavement Surface Characteristics of Roads and Airports, Toronto, Canada.

[3] Persson B N J (2001). Theory of rubber friction and contact mechanics. Chemical Physics, 115(8):3480-3860.

[4] Persson B N J, Tartaglino U, Albohr O, et al (2004). Sealing is at the origin of rubber slipping on wet roads. Nature Materials, (3):882-885.

[5] P. Cao, X.Q.Bai (2009). Study on the Relationship between Fractal Characteristics and Anti-sliding Performance of Asphalt Pavement. Lubrication And Sealing,(3):9-11.

[6] L.G.Zhang.,Z.D.Qian.,L.Q.Yang.el al (2013). Fractal expression of pavement morphology and anti-sliding performance of green concrete pavement. Highway. (5):85-88.

[7] T.J.Ji (2004). Influence of rainfall on the adhesion coefficient between tire and pavement. Southeast University.

[8] L.Cong, X.M.Huang (2011). Influence of Pavement Trictional Properties on Braking Distance. Journal of Southeast University: English Edition, (2): 210-212.

[9] ONG G P, FWA T F (2007). Wet-pavement hydroplaning risk and skid resistance: modeling. Journal of Transportation Engineering, 133(10): 590-598.

[10] FWA T F, ONG G P (2008). Wet-pavement hydroplaning risk and skid resistance analysis. Journal of Transportation Engineering,134(5): 182-190.

Acknowledgements

The research work described herein was funded by the Fundamental Research Funds for the Central research institutes (Grant No. 2018-A0028 & 2018-9013). This financial support is gratefully acknowledged.

IOP Conf. Series: Materials Science and Engineering **627** (2019) 012022 doi:10.1088/1757-899X/627/1/012022

Automatic Deformation Measurement Equipment for PZT Wafer

T Q Ren[1,2], X Li[2], Z F Lou[1,2] and X D Wang[1,2]

[1] Key Laboratory for Precision & Non-traditional Machining of the Ministry of Education, Dalian University of Technology, Dalian, China
[2] Key Laboratory for Micro/Nano Technology and System of Liaoning Province, Dalian University of Technology, Dalian, China

xdwang@dlut.edu.cn

Abstract. PZT wafers are the core driving parts to adjust the laser resonant cavity length of laser gyro. Usually, the PZT wafers are used in pairs, and the paired PZT wafers need to have close piezoelectric coefficient. To handle the pairing and screening of PZT wafers, an automatic deformation measuring equipment is developed by using a cartesian-coordinate robot frame, in which multiple photoelectric sensors are used to detect the key motion position automatically. Besides, vacuum absorption technology is also used in picking, carrying and placing PZT wafers to protect them from accidental injury. When driving voltage is applied to PZT wafer, the resulting micro displacement is measured by dual opposite inductive probes with relative measurement principle. This measuring strategy eliminates the influence of placement error of PZT wafers on the final measuring result. Compared with the existing manual measurement, the efficiency can be improved by 60%. The experimental results show that the equipment has high reliability and consistency. The measurement accuracy in full scale is no more than 0.5 μm and the repeat accuracy is superior to 0.1 μm.

1. Introduction

Laser gyro is an optical inertial device that measures the rotation velocity of object. It is widely used in aerospace field [1-2]. The working principle of laser gyro is to detect the optical path difference of two reverse laser beams in closed annular optical path, thereby to calculate the angular velocity of rotation[3]. Laser gyro needs to work at stable frequency. But its laser resonant cavity length usually varies in micron scale due to the influence of environmental factor, such as temperature and mechanical vibration. This leads to frequency variations which has fatal impact on working stability of laser gyro[4-5]. Therefore, laser gyro contains a displacement compensator to adjust the laser resonant cavity length so as to stabilize the working frequency [6]. Usually, the displacement compensator works based on principle of inverse piezoelectric effect, at this point, the PZT wafer acts as the core driving part [7-8]. The PZT wafer needs to be used in pairs, and the paired PZT wafers need to have close deformations at the same driving voltage. Therefore, the piezoelectric property, referring in particular to the deformation of PZT wafer, should be measured and matched before using.

Currently, manual measurement method is usually employed in production field. The manual measurement is inefficient and hard to guarantee the testing quality due to different workers' operation. This seriously affects the follow-up assembly work of laser gyro. Therefore, it is necessary to develop an automatic measuring equipment to deal with in-site batch measurement of PZT wafers with high efficiency and accuracy. For micron-sized deformation measurement of PZT wafer, optical and electrical methods are usually used. Optical method mainly refers to laser interferometry measurement, employing single-beam or double-beam laser interferometer [9-10]. The laser interferometry measurement can reach resolution of 10^{-14} m. However, optical measuring system is generally sensitive to the working condition [11]. And complex optical components for interferometry make the system ponderous and expensive. Moreover, it requires high technical skills for operators. Therefore, the laser interferometry measurement system is mainly used in laboratory for small batch and high precision measurement tasks. The electrical methods mainly include capacitor method [12] and inductance method [13]. For annular PZT wafer, the testing surface is narrow arc surface. In this case, it is easy to produce edge effect by using capacitance sensor. Comparatively speaking, the inductance method has low requirement on the testing surface and moderate precision. Moreover, it is ease for implementation and system integration to cooperate with automated production and detection. Therefore, inductance method has a much wider applicability for the advantages of simple operation, stable performance and high reliability.

To sum up, in order to deal with in-site batch measurement of PZT wafers with high efficiency, an automatic measuring equipment for PZT wafer of laser gyro is developed by using a cartesian-coordinate robot frame and dual opposite inductive probes with relative measurement principle. It is designed to meet PZT wafer with annular/rectangular shapes and multiple sizes. What's more, it has high reliability, consistency, and measurement efficiency.

2. Overall design of the equipment

As shown in figure 1, the PZT wafers are annular or rectangular with thickness of 0.4mm. According to the user requirement, the design objectives of the equipment are as follows: The measurement accuracy in full scale is no more than 0.5 μm and the repeat accuracy is superior to 0.1 μm. The measuring force is less than 0.2 N and the driving voltage range is DC 0~300 V. It can complete the task of automatic loading, measuring and sorting operations of 50 PZT wafers in a single batch.

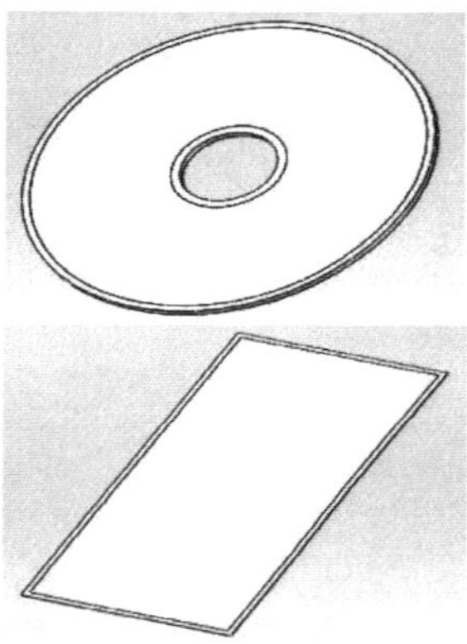

Figure 1. The PZT wafers to be test

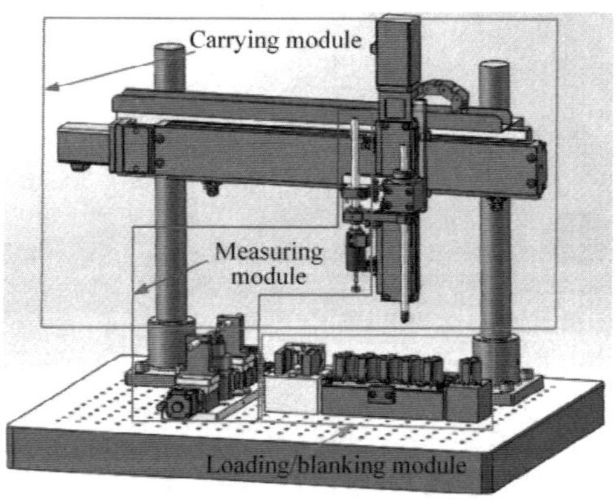

Figure 2. Overall design scheme of this equipment

The working process can be divided into three steps: loading/blanking, measuring, and sorting. Concretely, the loading process includes picking, carrying and placing PZT wafers. In order to avoid damage on PZT wafers during loading process, the vacuum adsorption picking mode is adopted. The cartesian-coordinate robot frame is used to locate the PZT wafer accurately. The measuring process includes accurate voltage loading and micron-sized deformation measurement. This equipment is used for in-site batch measurement of PZT wafers, which requires high reliability, operability and appropriate accuracy simultaneously. Therefore, inductance method is preferred, and dual opposite inductive probes with relative measurement principle is employed to reduce the requirements for the positioning accuracy. The accuracy of the driving power supply directly affects the measurement accuracy. So dedicated power supply for PZT with high control accuracy and stability is used. The sorting process is to match PZT wafers with similar piezoelectric property based on measuring results. This can be achieved through carrying PZT wafer to different recovery bins after measurement.

Based on the above analysis, three main functional modules are designed, i.e., loading/blanking module, carrying module and measuring module. The overall design scheme is shown in figure 2.

3. Equipment construction and working principle

The equipment is installed on an optical platform and multiple rubber vibration isolation cushions are set up under the platform to achieve good stability.

3.1. The loading/blanking module

The loading/blanking module consists of a feeding table and multiple recovery bins, as shown in figure 3. The feeding table mainly consists of an air cylinder and centring clamping jaw. Before measurement, the PZT wafers are stacked on the feeding table with relative freedom. Then the cylinder acts to drive the clamping jaw to tighten inward. Thus central alignment of the PZT wafers are completed.

After measurement, the PZT wafers are sorted based on measuring results. To avoid the stack of congeneric PZT wafers losing balance and toppling over, different type baffles are mounted on a special mounting plate to construct recovery bins. The mounting plate is designed as an independent part to facilitate disassembly for PZT wafers recycling. There are total six recovery bins, and one of them is for unqualified PZT wafer.

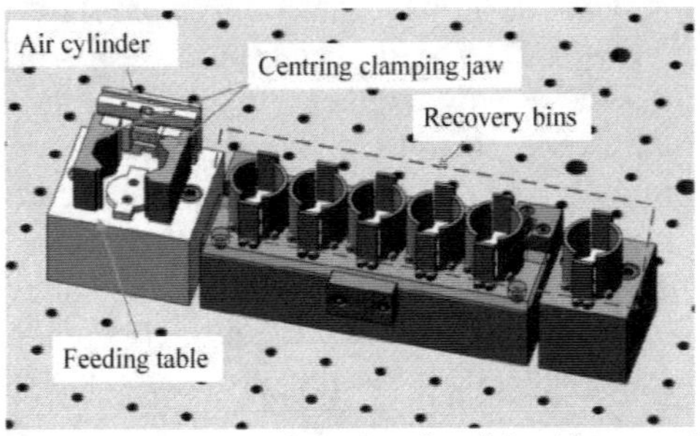

Figure 3. Structure of the loading/blanking module

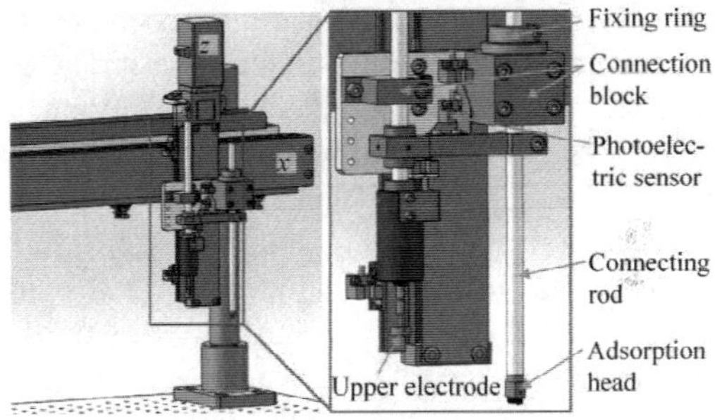

Figure 4. Structure of the carrying module

3.2. The carrying module

The carrying module completes the picking, carrying and placing operation of PZT wafers. It adopts planer type structure that consists of x-axis and z-axis, as show in figure 4.

The x-axis and z-axis are used for horizontal and vertical motion to pick/place the PZT wafer. The picking arm and the upper electrode are parallelly fixed on z-axis by connection blocks, and they are also fixed together via a common mounting beam. The mounting beam is separate from z-axis. The z-axis is equipped with photoelectric sensors, and the upper electrode and mounting beam are equipped with respective light barriers. During picking operation, when the picking arm is in contact with the PZT wafer, there will be relative motion between the mounting beam and z-axis due to the linear bearing embedded in right connection block. When the theoretical position is achieved, the light barrier triggers the corresponding photoelectric sensor to generate stop signal for z-axis. If this stop signal is responded incorrectly by accidents, z-axis will move on and the light barrier mounted on mounting beam triggers the top photoelectric sensor. By this time, an emergency stop signal is generated, which stop x-axis and z-axis simultaneously through interruption control. This protection strategy avoids damage on equipment and PZT wafer due to excessive motion of z-axis. The picking arm adopts double vacuum chucks to ensure stable picking operation. The vacuum chucks are mounted at the bottom end of the hollow connecting rod. There is a fixing ring at the upper part of the connecting rod to prevent it from falling off due to gravity.

The relative measurement using dual opposite inductive probes eliminates the influence of placement error of PZT wafers on the final measuring result. Therefore, the motion accuracy of the picking module is not required to be very high, but the travel range should be considered carefully.

3.3. The measuring module

The measuring module completes the voltage loading and micro-deformation measurement of PZT wafer. It mainly consists of measuring platform, inductive probes, and the upper and lower electrodes, as shown in figure 5 and figure 6.

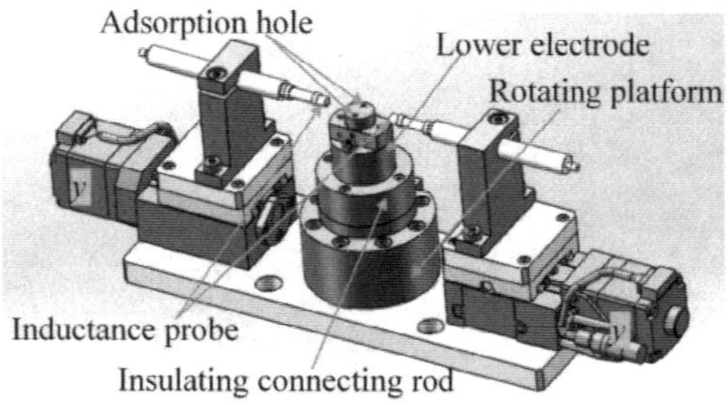

Figure 5. Structure of the measuring module

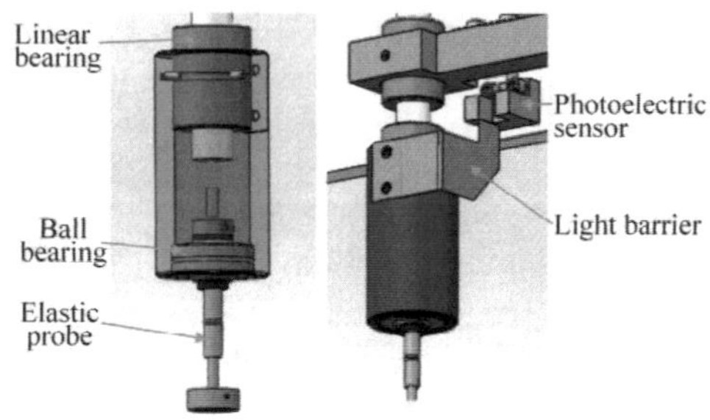

Figure 6. Structure of the upper electrode

An electric rotating platform is used as the main part of measuring platform, so as to measure the displacement of PZT wafer in different directions. Two inductive probes are mounted on two y-axes locating on opposite sides of the measuring platform. The lower electrode is integrated to the measuring platform, while the upper electrode is integrated to z-axis as shown in figure 4. During measurement operation, the PZT wafer is placed on the lower electrode by carrying module. Then the picking arm moves away and the upper electrode moves down to impact the PZT wafer. The two y-axes drive inductive probes moving towards to get in touch with PZT wafer. The y-axes stop until both two inductive probes have a certain amount of compression. Finally, the driving voltage is loaded through the upper and lower electrodes. PZT wafer is adsorbed on the lower electrode through two adsorption holes after placement. This vacuum adsorption works until the upper electrode completes the impaction operation. The upper electrode is designed as the combination of an elastic probe and a spherical bearing, as shown in figure 6. This structure makes the upper electrode has automatic attitude adjustment function, so as to overcome the problem of local force and incomplete contact. The upper electrode is equipped with a light barrier on its sleeve. Thus the principle of position control and hardware protection are similar to that of picking arm.

Likewise, the relative measurement principle reduces the requirements for accuracy of the moving platform in measuring module. The key parts are the inductive probes which requires low measuring force and high accuracy. In practice, inductive probes of GTL222-A from TESA SA is employed.

The stability of the driving voltage directly affects the final measuring results. It is necessary to select a driving power supply with high control precision. In addition, it must be programmable to achieve automatic operation. Finally, dedicated power supply for PZT, XE501-C from Harbin Core Tomorrow Science & Technology Co., Ltd., is selected. The most important parameters are Output voltage range, voltage stability and output ripple. They are 0~300 V, $<$0.1%F.S. /8hours and 5±20% mV, respectively. Usually, the piezoelectric property of PZT wafer characterized by deformation-voltage characteristic curve can be considered as linear. The user provides that the deformation of PZT wafer to be test is less than 0.02 μm with each voltage change of 1 V. Based on table 3, the maximum deformation of PZT wafer is about 0.02 μm/V×300 V=6 μm, and the voltage fluctuation at full-scale output is about 0.3 V, which will cause measuring error of 0.006 μm << 0.5 μm.

4. Measurement experiment and data analysis

The physical equipment is shown in figure 7. In order to check the performances of this equipment, an experiment using real annular PZT wafers was carried out.

Figure 7. The physical deformation measurement equipment

Firstly, three arbitrary PZT was selected and measured for 10 times at the same voltage of 100 V. The measuring results are summarized in table 1. It can be seen that the extreme deviation of measuring result for the same PZT wafer is 0.07 μm < 0.1 μm. This demonstrates that the repeat accuracy of the equipment meets the design objective.

Table 1. Experiment data for repeat accuracy.

Test serial number	PZT wafer1 (μm)	PZT wafer2 (μm)	PZT wafer3 (μm)
1	1.61	1.56	1.48
2	1.63	1.50	1.47
3	1.63	1.52	1.49
4	1.63	1.57	1.47
5	1.63	1.55	1.47
6	1.64	1.51	1.46
7	1.65	1.55	1.43
8	1.65	1.50	1.50
9	1.66	1.53	1.47
10	1.67	1.55	1.47

Subsequently, the same batch of 50 PZT wafers were measured by this equipment and existing manual measurement method at 100 V, respectively. The measurement processes cost 2 min and 5min, respectively, which means the work efficiency increases by 60%. Table 2 shows part of measuring results. After data processing, the average value of this equipment is 1.454 μm with standard deviation of 0.032 μm, and the average value of manual measurement is 1.451 μm with standard deviation of 0.084 μm, respectively. The dispersion degree of measuring result of this equipment is smaller, which indicates better reliability and consistency.

Table 2. Part of measuring results for batch measurement.

Test serial number	Equipment measurement data (μ m)	Manual measurement data (μ m)
1	1.45	1.48
2	1.47	1.47
3	1.46	1.52
4	1.49	1.38
5	1.49	1.47
6	1.46	1.44
7	1.46	1.52
8	1.45	1.48
9	1.50	1.36
10	1.41	1.43
...	...	...

In this work, the maximum deformation of PZT wafer is about 6 μm as mentioned in the last paragraph of section 3.3. Thus the maximum permissible error of the inductive probe is $0.2+2.4\times6^{-6}\approx$

0.2 μm. The measurement accuracy can be approximated as 2 times of the resolution of the inductive probe due to relative measurement principle. Therefore, the final measurement accuracy is much better than the design objective.

5. Conclusion

Based on comparative analysis of the existing techniques of deformation measurement for PZT wafer, a dedicated deformation measurement equipment is developed in this work. This equipment adopts cartesian-coordinate robot frame and two inductive probes with relative measurement principle. It realizes automatic operation of loading/blanking, measuring and sorting for PZT wafer, at the same time, it has self-protection function. Compared with the existing manual measurement in production field, the measurement efficiency can be improved by 60%, and the repeat accuracy is less than 0.1 μm. Obviously, it has higher reliability and consistency, which provides an efficient method for massive pairing application for PZT wafers. At present, the equipment is designed to meet only annular and rectangular wafer. The application scope of the equipment can be further expanded by modifying the measuring module and loading/blanking module.

6. References

[1] Passaro V M N, Cuccovillo A, Vaiani L, De Carlo M and Campanella C E 2017 *Sensors* **17** 2284

[2] Schwartz S, Gutty F, Feugnet G, Pocholle J P and Desilles G 2017 *Proc. Int. Conf. on Space Optics (Ajaccio, Corsica)* vol 10564 (SPIE) p 105643I-1–4

[3] Kuznetsov A G, Molchanov A V, Chirkin M V and Izmailov E A 2015 **45** 78–88

[4] Ding J C, Zhang J, Huang W Q and Chen S 2014 *Sensors* **14** 18711–27

[5] Wei G, Li G, Wu Y and Long X W 2011 *Meas J Int Meas Confed* **44** 1898–1903

[6] Zapotyl'ko N R, Katkov A A and Nedzvetskaya A A 2011 *J Opt Technol* **78** 644–5

[7] Wu B, Wang H F, Huang D D and Ding P M 2010 *Appl. Mech. Mater.* **29-32**, 1640–45

[8] Su S Q and Li J M 2013 *Nanotechnology and precision engineering* **11**, 551–6

[9] Wu H, He L B, Zhou J L and Yang P 2015 *Key Eng Mat* **645**, 920–5

[10] Schwartz S, Gutty F, Feugnet G, Pocholle J P and Desilles G 2017 *Proc. Int. Conf. on Space Optics (Ajaccio, Corsica)* vol 10564 (SPIE) p 105643I-1–4

[11] Gamboa B M, Malladi M, Vadlamani R, Guo R Y and Bhalla A 2016 *Proceedings of SPIE - Photonic Fiber and Crystal Devices: Advances in Materials and Innovations in Device Applications X (San Diego, CA)* vol 9958 (SPIE) p 99581E-1–7

[12] Steinhausen R, Kerna S, Pientschke C, Beige H, Clemens F and Heiber J 2010 *J. Eur. Ceram. Soc.* **30**, 205–9

[13] Cheng C H, Chang M C and Liu W F 2011 *Guangxue Jingmi Gongcheng* **19**, 2255–62

Acknowledgments

This research work was supported by Major Project of Basic Scientific Research of Chinese Ministry (Grant No. JCYK 2016 205 A003), National Science and Technology Major Project of China (Grant No. 2013ZX04001091), and the National Natural Science Foundation of China (No. 51621064).

Automatic Hot Pressing System for Indium Packaging

Yi Luo, Yawei Li, Xiaodong Wang*and Xiaoxu Qiao

Key Laboratory for Micro/Nano Technology and System of Liaoning Province,

Dalian University of Technology, Dalian 116024, China

* Corresponding author: xdwang@dlut.edu.cn

Abstract. Hot pressing is one kind of effective assembly technology to join two parts together. Automatic hot pressing need control both temperature and pressing force precisely and programmable. In this paper, an automatic hot pressing system for indium packaging is presented, the design motivation is to made the system reliable for application in industry and reach the accuracy requirement. A screw jack with servomotor is selected to actuate the hot pressing head to generate the pressing force, a linear encoder is used to detect the displacement of the screw jack to improve its motion accuracy. A force sensor is mounted underneath the 2-D platform to feedback the pressing force in real time. The hot pressing head is made up of four heating rods and a temperature sensor to control the pressing temperature. The exchangeable manipulator is designed to pick up parts with different size and shape. The machine vision is employed to calibrate the first set of parts to make the system flexible to product variety. The experimental results shown that the maximum force error is 9 N at working load of 2000 N and the temperature error is less than ± 0.3°C. The cycle time is 630 s.

1. Introduction

Assembly is the capstone process in manufacturing [1]. Currently, many efforts are made to change assembly processes from manual to automatic.

Hot embossing or hot pressing is one kind of assembly technology to join two parts together. Automatic hot pressing need control both temperature and pressing force precisely and programmable. Kurita [2] proposed laser embossing complex machining for glass, and experimentally demonstrated by changing the laser irradiation conditions used to heat a mould. Lee [3] developed a hot press machine for manufacturing complex contoured components using super plastic forming and diffusion bonding technology. The maximum working height is 1500 mm and platen size is 1700×1700 mm^2. The capacity is 200 ton, and the maximum operating temperature is 1200°C. Chen [4] reported a press device used for hot embossing and thermal bonding of PMMA micro fluidic chips which is suitable for the fabrication of PMMA microchips. Hale[5] developed a hot embossing system for manufacturing micro scale parts, the layout of the frame is a welded steel press frame in an arch shape, The actuator is a pneumatic cylinder which can produce a maximum force of 9800 N. The heating components are two ceramic heaters. The minimum cycle time is two minutes, and the maximum embossing area is 25mm by 75mm.

Hot embossing can joining two parts made of same or different materials, and the interface material should relieve the mechanical stress and absorb the strain cause by mismatch of coefficient of thermal expansion. Indium with the low melting point of 157°C is a good candidate for interface material [6]. It is also a good choice for cryogenic joint (below -55°C) [7]. However, indium also has one serious drawback. The bonding quality drops dramatically when indium is oxide. So place the indium to the right place automatically and pressing immediately is critical to the bonding quality. However, there are not commercialized hot pressing device for indium packaging yet.

In this paper, an automatic hot pressing system was designed. Indium was selected as the interface material to hermetic package parts made of glass and aluminum alloy. Machine vision was employed to calibrate the first set of parts to make the system flexible to product variety. Screw jack with servomotor was used to actuate the pressing head to apply pressing force. Four heating rods and a temperature sensor was selected to control the pressing temperature.

2. Indium packaging system and working principle

2.1 The diagram of the hot pressing system

The designed automatic hot pressing system is used to realize hermetic packaging between an aluminum part and a glass base with indium as interface material. The requirements for the bonding system are: bonding temperature is range from room temperature to 150°C with the accuracy of ±2°C, the maximum pressing force is 1500 N with the control accuracy higher than ±10 N. To meet the requirements, the design diagram of the pressing system is shown in figure 1.

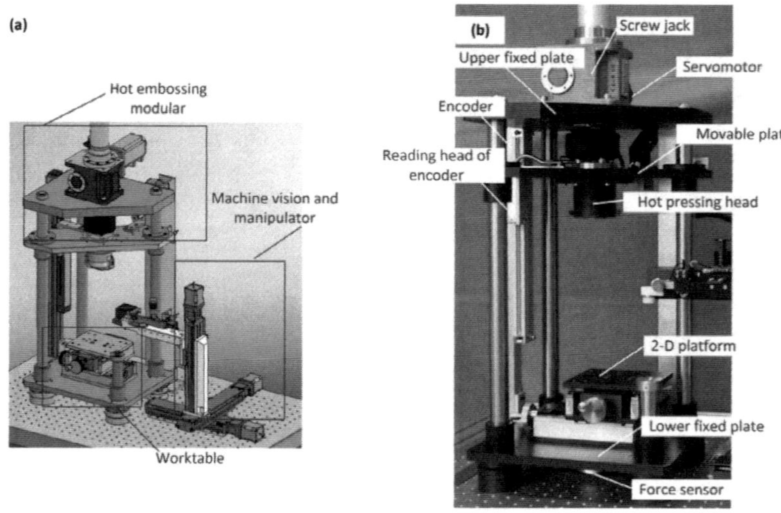

Figure 1. The automatic hot pressing system for indium packaging (a) the diagram and (b) the setup of the hot pressing module

The hot pressing system is made up of three modular. The hot pressing modular is the key component to generator pressing force and pressing temperature precisely. The machine vision and manipulator is consisted of machine vision unit and exchangeable manipulator unit. The CCD camera and the manipulator are fix together and actuated by an X-Y-Z linear stage. The worktable is a 2-D platform to support the glass base part, which can be adjusted in X and Y directions.

The indium packaging procedure is as follows: Firstly, the glass base is placed on the two-dimensional platform, then move the CCD camera to get a clear image of the hole on the glass base, record the position information of the X-Y-Z linear stage. Secondly, take the indium ring from alcohol, placed on the glass base and made the hole and ring concentric, follows by picking the aluminum part by the manipulator, the computer calculate the its displacement according to the position of the CCD camera to make the aluminum part align to the glass base part correctly. Finally, the hot pressing head is heated to the pre-set temperature, the screw jack moves down to apply the pressing force for a period of time, the hermetic packaging between aluminum part and glass base thus realized.

2.2 Pressing force applying strategy

Pressing force is the key parameter to get successful indium packaging. Servomotor drive has the advantage of high precision and easy to regulate the speed. In the designed pressing system, pressing force is applied and controlled indirectly by the hot pressing head which is actuated by the servomotor. That is, the force is controlled by the displacement of the hot pressing head and system stiffness, the pressing force applying strategy is shown in figure 2.

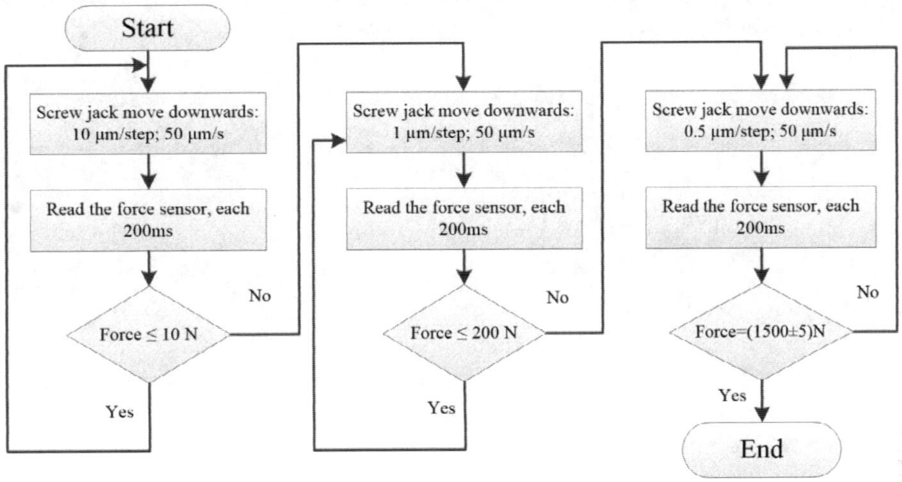

Figure 2. Screw jack speed and displacement combination control for pressing force applying

The servomotor and screw jack is mounted on the upper fixed plate to actuate the hot pressing head, while the latter is mounted on the movable plate to exert temperature and pressing force to the parts. Since the displacement of the hot pressing head is a crucial to packaging quality, to reach the close loop control, a linear encoder is integrated in this system. The read head of the encoder is fixed to the movable plate to get the displacement information.

The 2-D platform is mounted on the lower fixed plate to carry the parts. A force sensor is located underneath the lower fixed plate to get and feedback the force information in real time. The hot pressing head is move at a constant speed when it does not contacted the part. When the hot pressing head reaches the part, the control method changed to the stiffness-displacement method.

$$F = \int_0^d K_{eff} \cdot \Delta s \quad (1)$$

Where F is the pressing force exerted by the hot pressing head; K_{eff} is the equivalent coefficient of elasticity, it is depends on the system material and structure, can be measured by experiment; Δs is the small displacement increment of the hot pressing head.

Experiments shown that in this system, when the screw jack actuated hot pressing head is steadily contacted the part, 0.5 μm increment in the displacement results to about 3 N in pressing force.

2.3 Temperature control of the hot pressing head

The detailed structure of the hot pressing head is shown in figure 3. The main components are four heating rods and one temperature sensor, which are insert into the hole in the hot pressing body and fixed by the thermal conductive silica glue. The power of each heating rod is 44 W. The temperature sensor is a platinum resistance thermometer, the working temperature range from -70°C to 300°C with the accuracy class of 1/3B.

2.4 The exchangeable manipulator

If the parts with different size and dimension need to be assemblies, different manipulator should be easily changes. The exchangeable manipulator is shown in figure 4. Different tools insert into the tool change hole on the manipulator and fixes the screw, thus the tool exchange completes.

2.5 Machine vision

Machine vision was employed to calibrate the first set of parts to make the system flexible to product variety. Meanwhile, it also solves the problem of visual occlusion in alignment if the part was opaque. Put the glass base on the 2-D platform, turn on the CCD camera and adjust the 2-D platform to make the assembly mark on the glass base at the center in the image. The green line is the scale bar and the red circle is the alignment mark, as shown in figure 5. Record the exact position of the X-Y-Z linear stage and add the offset of the CCD camera and the tool, the manipulator then knows how to move the part to the right position.

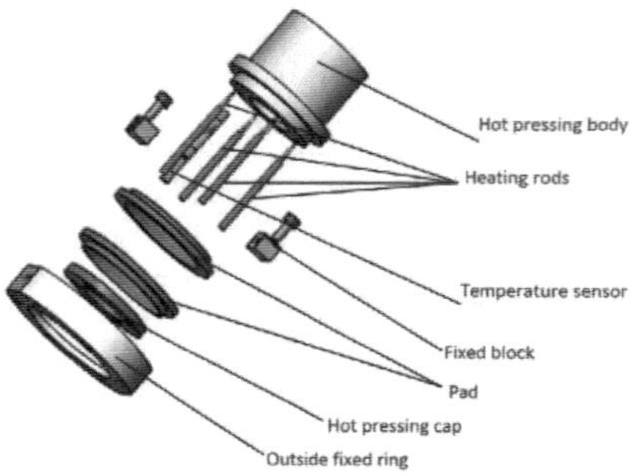

Figure 3. The explosive view of hot pressing head

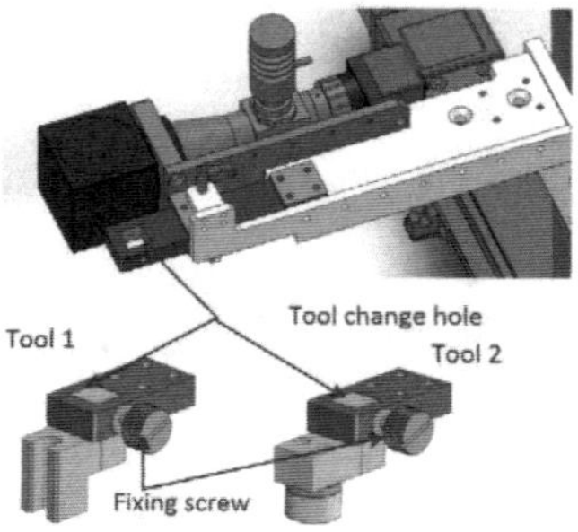

Figure 4. The exchangeable manipulator

3. Experiments results and discussion

3.1 The pressing force control

The pressing force is the important parameter in indium packaging quality control. Below the pre-set value may result to poor sealing performance, while large pressing force potentially harmful to the aluminum part and the glass base.

Standard force sensor was selected to measure the pressing force simultaneously with the force sensor of the system. The standard force sensor's working range is from 0 to 5000 N, the repeatability is 0.02%FS. It is place on the 2-D platform and the two force sensor are coaxial. Moving the hot pressing head downwards to apply force and read the values of the two sensor, the experiments repeated seven times and the data listed in Table 1.

Table 1. Data of pressing force in the accuracy testing experiments

The system value（N）	The value of the standard force sensor（N）							Max. error (N)
	1	2	3	4	5	6	7	
100	101	101	103	101	101	101	101	3
500	507	508	508	508	507	507	508	8
1000	1004	1006	1002	1006	1001	1006	1004	6
1500	1507	1507	1505	1506	1507	1508	1505	8
2000	2009	2006	2007	2008	2006	2002	2006	9

According to Table 1, the maximum error is 9 N at working load of 2000 N and can meet the design requirement, which is ±10 N at working load of 1500 N.

The fluctuation of the real pressing force may be caused by the control system, it read the force sensor each 200 ms, and calculating and feeding back the signal also need time. Reduce the reading interval may benefit the force accuracy, however it also put forward stricter demand for computer and algorithm. So concerned of economy and stability, the force control parameters were selected.

3.2 The temperature control of the hot pressing head

The temperature at the surface of the hot pressing head also influence the sealing quality. The temperature testing results is listed in Table 2. The temperature error is less than ± 0.3°C, much less than the equipment of ± 2°C. The aluminum part and the indium ring have small thermal capacity, glass base is not a good conductor, so the temperature will not have a large fluctuation when the hot

pressing head just contacts the part, the heating power also sufficient to make up the heat loss in the force holding process.
In indium packing experiment, the temperature is controlled at 120°C constantly, the pressing force is ramped increased to 900 N, and the results is shown in figure 6. The cycle time for each packaging is 630 s, including 300 s as hold time required by the packaging technique.

Table 2. Temperature of the sensor and the thermometer after compensation

Sensor temp. (°C)	30	50	70	90	110	130	150
Thermometer temp. (°C)	30.2	49.8	70.1	89.7	109.8	129.9	150.2
Deviation (°C)	0.2	-0.2	0.1	-0.3	-0.2	-0.1	0.2

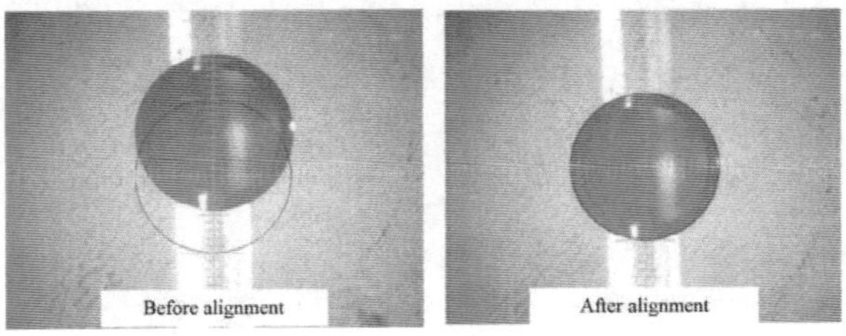

Figure 5. The alignment of the parts

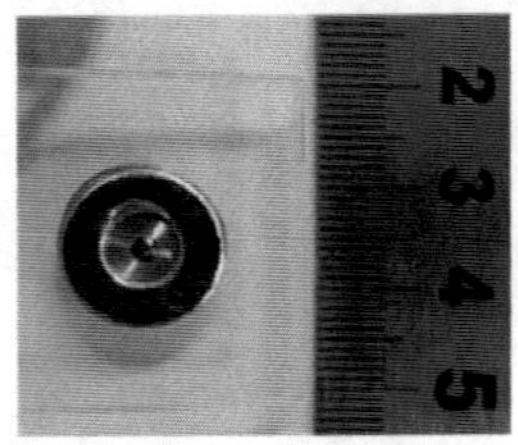

Figure 6. The packaging result

4. Conclusions
An automatic hot pressing system for indium packaging is presented in this paper. The motivation is to make the system reliable for industry application, meanwhile meet the design requirements. The pressing force is generated by screw jack with servomotor. The force sensor is mounted underneath the 2-D platform to get the pressing force in real time. The hot pressing head is made up of four heating rods and the temperature sensor to control the pressing temperature. The exchangeable manipulator is designed to pick up parts with different size and shape. The machine vision is employed to calibrate the first set of parts. The maximum force error is 9 N at working load of 2000 N and the temperature error is less than ±0.3°C. The cycle time of packing is 630 s.
Future work will be focus on study the sealing quality of the indium and adjust the pressing parameters.

5. Reference
[1] Whitney D E 2004 *Mechanical Assemblies* (Oxford: Oxford University Press)
[2] Kurita T, Ogura I and Ashida K 2018 *J. Mater. Process Technol.* **254** 248-253
[3] Lee H S, Yoon J H, Yoo J T and Yi Y M 2011 *Adv. Mater. Res.* **228-229** 563-567
[4] Chen Z, Zhang L Y and Chen G 2010 *Electrophoresis* **31** 2512-2519
[5] Hale M, Hardt D E 2009 *ASME 2009 Int. Mechanical Engineering Congress and Exposition (Lake Buena Vista)* vol 12 (New York: Amer SOC Mechanical Engineers) p455
[6] Qin Y, Flajslik K , Sherzer B, et al. 2016 *66th IEEE Electronic Components and Technology Conf. (Las Vegas)* (Los Alamitos:IEEE Computer SOC Los Alamitos) p2151
[7] Chang R W and McCluskey F P 2009 *Cryogenics* **49** 630-634

Acknowledgments
Authors wishing to acknowledge the financial support of Major Project of Basic Scientific Research of Chinese Ministry (Grant No. JCYK 2016 205 A003).

IOP Conf. Series: Materials Science and Engineering **627** (2019) 012024 doi:10.1088/1757-899X/627/1/012024

Molecular Dynamics Simulation of Heat Transfer at Contact Interface of Hot Forming Die

Baoshan Lu[1], Leigang Wang[1] and Xiang Ma[2*]

[1] School of Materials Science and Engineering, Jiangsu University, Zhenjiang 212013, China

[2] SINTEF Industry, POB 124, Blindern, 0314 Oslo, Norway

xiang.ma@sintef.no

Abstract. During metal hot forming process, the billet and the die interact by contact under high temperature, high pressure and severe sliding. In this paper, an atomistic approach is utilized to study the behaviour of heat transfer at the contact interface of hot forming die and the material. Based on non-equilibrium molecular dynamics and phonon scattering theory, the phonon heat transfer process at the billet/die rough interface at nano-scale is analysed. The effects of interfacial lattice defect concentration and average interfacial temperature on the phonon heat transfer coefficient are examined. It is found that when the lattice defect concentration increases the phonon mismatch at the interface deteriorates and the interfacial thermal resistance increases, which leads to a sharp decrease in phonon heat transfer coefficient. A higher temperature of the contact body and a higher average interface temperature lead to a higher probability of inelastic scattering of phonons at the interface. Moreover, the high frequency phonons at high temperature might be decomposed into two or more low frequency phonons during the energy transmission, which increases the propagation of phonon at the interface; thus the phonon heat transfer coefficient increases accordingly.

1. Introduction

The influence of incompletely contacted interface on heat transfer is an important topic in the field of micro-scale heat transfer. Molecular dynamics (MD) calculation is an effective method to carry out research on this subject to estimate some thermophysical properties which are difficult to be obtained experimentally. However, there has been few public reports on the heat transfer at the solid metal contact interfaces. There are two main reasons for this: (1) there is a lack of potential functions to precisely describe the atomic interactions of alloyed materials or dissimilar interfaces; (2) the MD method mainly studies the heat conduction problem caused by phonons, while the heat conduction in solid metal is dominated by free electrons where the phonon heat conduction only accounts for a small proportion of the global thermal conductivity [1]. It is generally believed that this small proportion of phonon thermal conductivity do not reflect the overall characteristics of solid metal interfacial heat transfer. These have become the bottleneck restricting the application of MD to the study of solid metal interfacial heat transfer.

However, for metals with lower thermal conductivity such as Ni, the thermal conduction caused by phonons cannot be ignored, especially at the micro/nano scale interface where the free electrons are limited by the surface and size effect leading to a sharp decrease in heat conduction [2]. The same could be true for the billet/die interface in hot forming. The rough or imperfect contact interface of the billet/die carries various non-metallic inclusions such as air gap, nitride layer, oxide layer or oil stain.

These non-metallic inclusions greatly inhibit the free electron conduction, while the phonon heat conduction is significantly enhanced and rapidly becomes the dominant factor in heat transfer. Therefore, this paper will use a non-equilibrium Molecular Dynamics method (NEMD, [3]) to examine the phonon heat transfer behaviour in an asymmetric billet/die interfacial model. The effects of the interfacial lattice defect and the average interfacial temperature will be studied.

2. F-S potential function and the NEMD model

In the MD simulation, the selection of the interaction potential between the particles in the system is very important as it determines the precision of the simulation results and the computational time. One effective potential energy model for the metallic atomic system is the multi-body potential such as the Finnis and Sinclair model (F-S) which is based on the tight-binding theory of metallic energy bands [4]. The F-S potential function is chosen in this paper for the MD calculation of a heterogeneous Fe-Ni atomic system, close to the hot forming application where the main alloy elements are Fe (in a typical H13 steel die) and Ni (in a GH4169 superalloy billet). The F-S potential parameters of Fe and Ni is obtained from the literature [4].

The NEMD method requires a temperature gradient between the atomic layers of the thermally conductive region. The thermostat method is used to apply thermal disturbance to the heat conduction zone along the thickness or Z directions of the hot zone (the billet) and the cold zone (the die) respectively [3]. If the statistical heat flow in the system is J_z and the interfacial temperature difference is ΔT, then the interfacial heat transfer coefficient h_{ph} (or phonon heat transfer coefficient) caused by phonon action is give as:

$$h_{ph} = J_z / \Delta T \qquad (1)$$

The relationship between the phonon heat transfer coefficient h_{ph} and the global interfacial heat transfer coefficient (IHTC) is that h_{ph} is the part of IHTC and accounts for a large proportion of IHTC values at the micro/nano-scale incomplete contact interface.

During the simulation, the atomic motion equation is numerically integrated using the velocity Verlet algorithm. The integration step is set to 4 *fs* and the total integration step is 3 million. The total simulation time t_s is 12 ns. Among them, the first 9*ns* is used for the initial balance of the system and the last 3*ns* is used to calculate the temperature of each atomic plane and the heat flow.

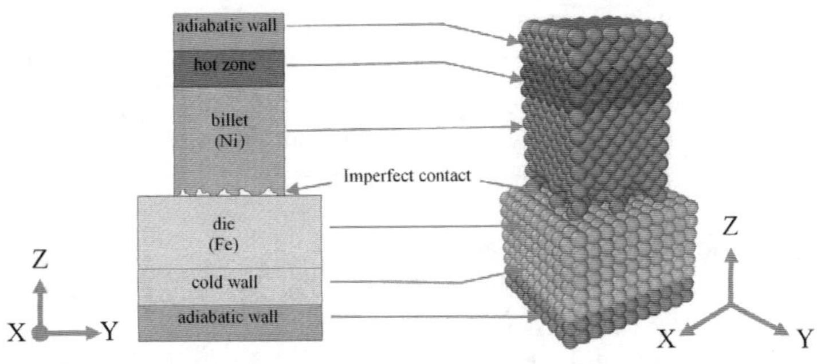

Figure 1. MD model of heat transfer at asymmetric interface

An asymmetric interfacial heat transfer MD model is constructed as shown in Figure 1, in which the periodic boundary conditions are applied in the X and Y directions. The superlattice composed of Ni atom (face-centered cubic structure) and Fe atom (body-centered cubic structure) forms an asymmetric contact interface, and the Ni atoms are randomly removed to indicate its incomplete contact with the surface of the Fe atoms. The cross-section (XY plane) of the model is square in which the Fe atomic layers are 10UC×10UC (UC is the unit cell) and the Ni atomic layers are 5UC×5UC. The thicknesses of the heat conduction zones of the Ni and Fe atomic layers are 6UC and 4UC, respectively. To prevent atoms on the surface of the hot and cold regions from escaping from the system, a layer of adiabatic wall with 2UC is added on the outer side of the hot and cold regions, respectively. The maximum temperature difference in the system is 100 K.

3. MD simulation results and discussions

3.1 Effect of interfacial lattice defect concentration on phonon heat transfer coefficient

A certain proportion of Ni atoms are randomly removed at the billet/die interface to simulate the effect of incomplete contact on the phonon heat transfer coefficient. The ratio of the number of Ni atoms deleted at the interface to the total number of Ni atoms in the heat transfer region indicates the interfacial lattice defect concentration. Figure 2 shows the temperature distribution along the Z direction when the pre-set temperatures of the hot and cold regions are 673 *K* and 573 *K*, respectively, and the interfacial lattice defect concentration of Ni is 5%. It can be clearly seen that the temperature in the vicinity of the hot and cold regions is nonlinear, while the temperature in the region far away from the hot and cold regions is approximately linear. The temperature in the heat conduction region of the Ni atom decreases rapidly. In the heat conduction zone of Fe atom, the heat resistance encountered is small when the heat flow flows through the Fe atomic layer, which is due to that the thermal conductivity of Fe atom is much higher than that of Ni atom. Thus, the temperature change in the Fe region is much smoother than in the Ni region. At the Ni/Fe contact interface, there is a significant temperature discontinuity due to the sudden lattice discontinuation, as high as 36.4 *K*.

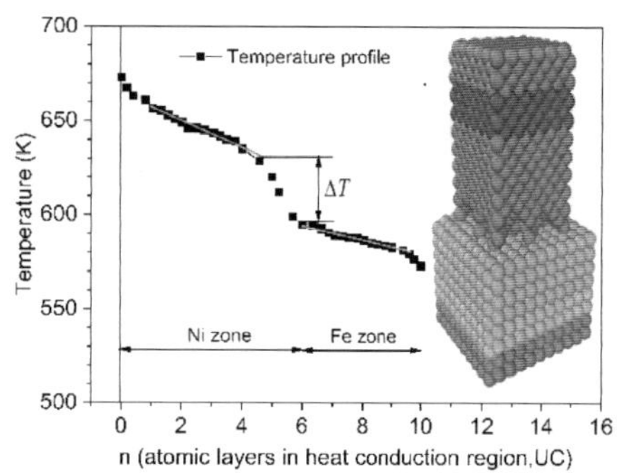

Figure 2. Temperature distribution of Ni/Fe atomic system.

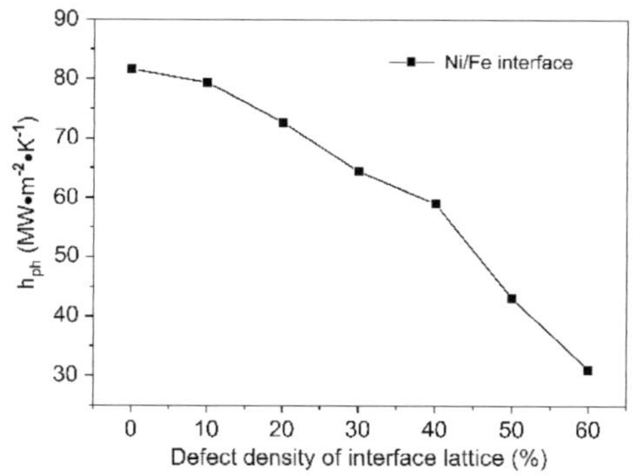

Figure 3. Variation of h_{ph} with defect density of interface lattice.

Figure 3 shows the relationship between h_{ph} and the interface lattice defect concentration when the preset temperatures of the hot and cold domains are 673 *K* and 573 *K*, respectively. As the interfacial lattice defect concentration increases, the interfacial thermal resistance increases and h_{ph} decreases accordingly. When the interfacial lattice defect concentration is less than 40%, h_{ph} decreases slowly with the increase in the interfacial lattice defect concentration. But when the defect concentration exceeds 40%, the phonon mismatch at the interface deteriorates, resulting in h_{ph} dropping rapidly.

To examine the effect of the phonon mismatch at the interface on the interfacial heat transfer, the density of phonon (DOS) of the Ni/Fe atom at the perfect contact interface is obtained at the average interfacial temperature of 623 *K*, as shown in Figure 4. The phonon frequency of Ni atom at the interface ranges from 0 to10 THz, while the phonon frequency of Fe atom is from 0 to 14.5 THz. The phonons of the Ni and Fe atoms at the interface remain matched in the frequency range of 0～3 THz and 8～10 THz, while the phonons have a significant mismatch in other frequency ranges.

When the interfacial lattice defect concentration is adjusted to 30%, the Ni and Fe atoms are in incomplete contact at the interface and the DOSs of the Ni/Fe atom at the interface are obtained at the average interfacial temperature (the arithmetic mean of the hot and cold zone temperatures) of 623 *K*, as shown in Figure 5. The removal of Ni atoms at contact surface causes the surface lattice defects, reducing the phonon binding probability on both sides of the interface and making the degree of acoustic mismatch worse. At the frequency range of 8-10THz, the phonons no longer match. However, the low-frequency phonons in the range of 0 to 3 THz are almost unaffected by the interfacial lattice defects throughout the heat transfer process. It shows that low-frequency phonon heat transfer is dominant in the interfacial heat transfer. The intermediate or high frequency phonons do not match at the interface, and the degree of mismatch increases with the increase in the interfacial lattice defect concentration which causes h_{ph} to decrease rapidly. Therefore, to improve the phonon interfacial heat transfer, it is necessary to improve the matching degree of low or intermediate frequency phonons at the interface.

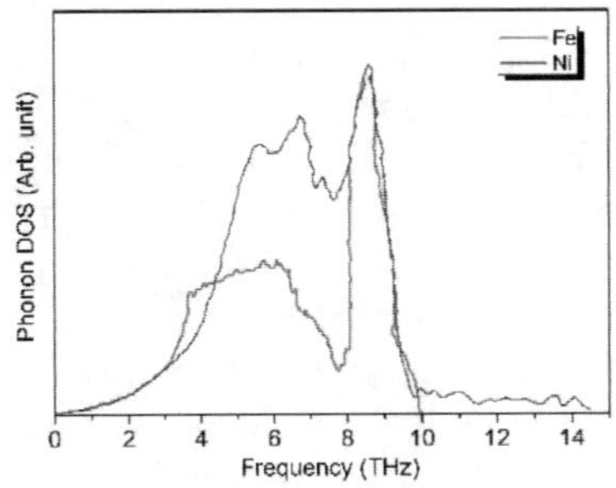

Figure 4. Density of phonon states of Ni and Fe at complete contact interface.

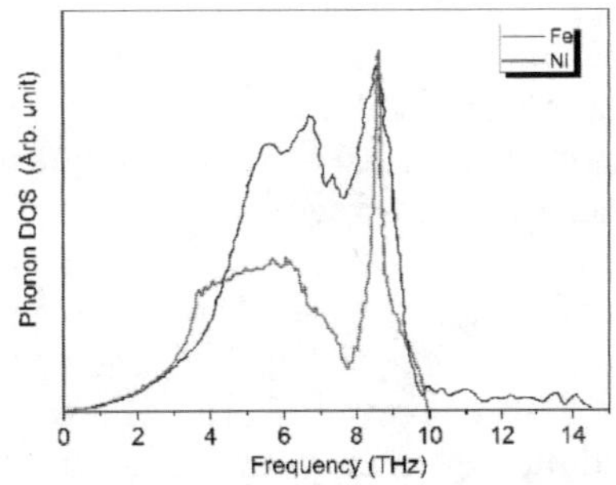

Figure 5. Density of phonon states of Ni and Fe at incomplete contact interface.

3.2.Influence of average temperature of different interfaces on phonon heat transfer coefficient

Figure 6 shows the phonon heat transfer coefficient obtained when the interfacial lattice defect concentration is 5%, the hot zone temperature remains fixed (673 *K*) and the cold zone temperature increases; or the cold zone temperature remains fixed (613 *K*) and the hot zone temperature increases. Here the temperatures of hot and cold zones correspond to the initial forging temperature and the preheating temperature of the billet and the die at the contact surface. When the average interfacial temperature continues to increase, h_{ph} also increases. h_{ph} increases fast at an early stage of heat conduction, then its growth rate is gradually slowed down. It is also observed that if the average temperature of the interface is the same, h_{ph} increases always greater when the temperature of the hot zone remains unchanged and the temperature of the cold zone increases, than that when the temperature of the cold zone remains unchanged and the temperature of the hot zone increases. This indicates that the phonon heat conduction is not sensitive to the temperature change in the hot zone, and the influence of the temperature change in the cold zone on the phonon heat transfer coefficient is more significant.

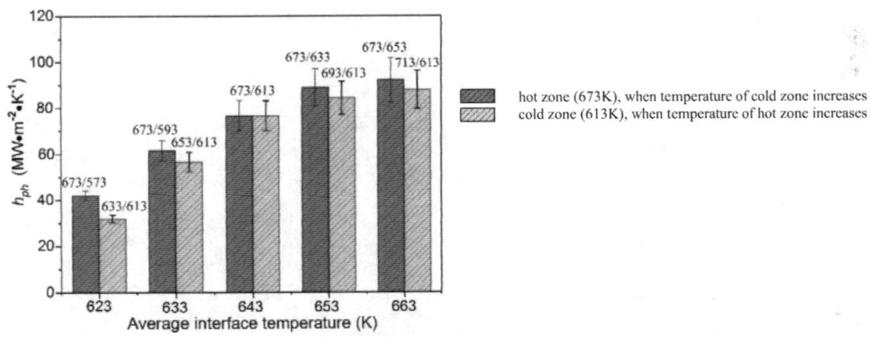

Figure 6. Variation of h_{ph} with different average interface temperature.

The maximum heat transfer coefficient h_{ph} calculated in Figure 6 is 90 MW/(m^{2}K), which is several orders of magnitude higher than the experimental results of the global interfacial heat transfer coefficient (IHTC, [5]). In fact, h_{ph} is part of the IHTC and should be less than the IHTC value. However, in the MD simulation of this example, there are three main reasons for the high phonon heat transfer coefficient h_{ph}: (1) when performing MD simulation, the thickness of the heat conductive layer is only a few nanometers. The surface roughness in the real contact interface during hot forming is at least micron-scale. (2) the non-metallic inclusions are neglected in the MD simulation which may have negative impact on the interfacial heat transfer. (3) the simplified model error and calculation error and their error accumulation.

To further investigate the effect of interfacial temperature changes on the phonon heat transfer coefficient h_{ph}, Figure 7 shows the phonon state density matching curves for Ni/Fe atoms at different average interfacial temperatures when the interfacial lattice defect concentration is 30%. When the average interfacial temperature increases from 623 K to 663 K, the low-frequency phonon densities of Ni and Fe atoms increase at the interface, and the low-frequency phonon matching degree also increases. The high-frequency phonons and their density peaks continue to decompose during the energy transmission process. As the frequency of high-frequency phonons is continuously attenuated, the frequencies of high-frequency phonons of Ni and Fe atoms are attenuated from 9.25 THz and 13.7 THz at 623 K to 8.7 THz and 10.5 THz at 663 K, respectively. The peaks of DOS of corresponding to high-frequency phonons also decrease with the increase in the interfacial temperature. At the same time, the intermediate and low frequency phonons increase continuously, so that their phonon density curves are shifted upwards. This shows that high-frequency phonons at high temperature are decomposed during energy transmission, and are decomposed into two or more low-frequency phonons. The latter promotes the matching of low-frequency phonons and increases the probability of phonon transmission at the interface. Therefore, the phonon heat transfer coefficient h_{ph} generally increases with the increase in the interfacial temperature, which explains the basic experimental fact that the IHTC increases with the increase in the interfacial temperature.

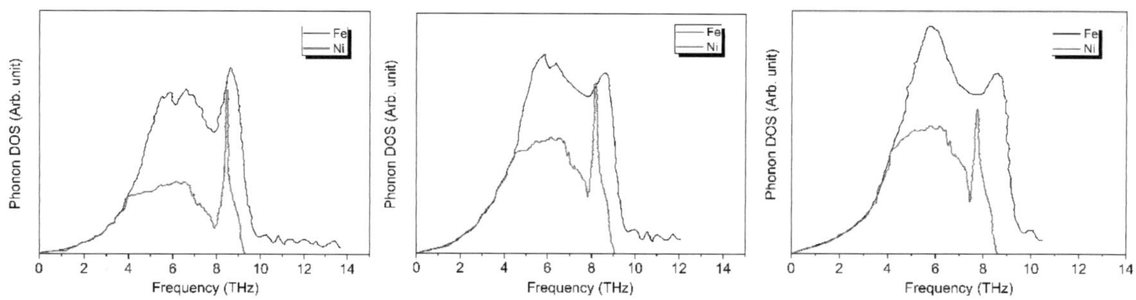

(a) 623K (b) 643K (c) 663K

Figure 7. Phonon DOS matching curves of Ni/Fe atom under different average interface temperatures.

4. Conclusions

By use of non-equilibrium molecular dynamics and phonon scattering theory, the phonon heat transfer process at the billet/die imperfect interface at nano-scale is analysed. It is found that when the lattice defect concentration increases, the phonon mismatch at the interface deteriorates and the interfacial thermal resistance increases, which leads to a sharp decrease in phonon heat transfer coefficient. A higher temperature of the contact body and a higher average interfacial temperature result in a higher probability of scattering of phonons at the interface. Moreover, the high frequency phonons at high temperature might be decomposed into low frequency phonons.

5. References

[1] Myers H P 1990 *Introductory Solid State Physics* (London: Taylor & Francis).
[2] Yuan S P and Jiang P Y 2005 *J. Engineering Thermophysics* **A1** 175
[3] Todd B D and Daivis P J 2017 *Nonequilibrium Molecular Dynamics: Theory, Algorithms and Applications* (Cambridge: Cambridge University Press)
[4] Finnis M W and Sinclair J E 1984 *Phil. Mag.* **50** 45
[5] Lu B, Wang L and Huang Y 2016 *Applied Thermal Engineering* **108** 516

Acknowledgement

This work is supported by the National Natural Science Foundation of China (project numbers: 51275216 and 51775249).

Laccase Enzyme Immobilization on Ferromagnetic Support Material for Oil Palm Empty Fruit Bunches (OPEFB) Pretreatment

Shah Samiur Rashid[1,*], Hu Shian Huan[1], Mohd Hasbi Bin Ab. Rahim[1], Md. Belal Hossain Sikder[1,2] and Ahmed Abdulsalam H Sasi[1]

[1]Faculty of Industrial Sciences & Technology, Universiti Malaysia Pahang, Gambang, Kuantan 26300, Pahang Darul Makmur, Malaysia
[2]Department of Food Engineering and Tea Technology, Shahjalal University of Science and Technology, Bangladesh

*Corresponding author: samiur@ump.edu.my

Abstract. Rapid development of oil palm industry causes a lot of environmental issues due to the large number of waste products such as OPEFB. OPEFB is rich in lignocellulose which can be hydrolyzed to product sustainable biofuel, an alternative sources of energy. The production of biofuel from lignocellulose waste requires pretreatment process to enhance efficiency of saccharification of the sugar content of the lignocellulose waste. Enzymatic pretreatment technique is more desirable as this technique does not generate greenhouse gas or any inhibitor that can interferes the saccharification process. Enzyme immobilization technique also involved to increase the stability of the enzyme and promote reutilization of the enzyme. In this study, laccase enzyme was immobilized on ferromagnetic particles to promote easy separation of the immobilized laccase enzymes from the reaction media. The result shows that laccase enzyme immobilized on the nickel particles which encapsulated with silica. 0.235 IU/g of immobilized laccase degraded 58 % of the lignin content in 0.5 g of OPEFB after 24 hours. Pretreated OPEFB increase the efficiency of saccharification by 161% in compare with the untreated OPEFB.

1. Introduction

Climate change due to global warming is considered one of the major negative impacts of the modern industrial civilization that believed to be contributed through emission of anthropogenic greenhouse gas (GHG) because of excessive consumption of fossil fuel as the main source of energy [1]. Therefore, intense urge has been perceived to find an alternative sustainable renewable energy. Natural biomass is considered as an alternative source of renewable energy due to its higher content of convertible sugar polymer, cellulose. There are a lot of biomasses such as waste paper, different fruit peels (banana, orange and pomegranate etc.), sugarcane bagasse, different natural fibers (jute, kenaf, OPEFB etc.) [2] have been explored as they are abundant in nature. However, due to many reasons there is no sustainable technology for alternative renewable energy production yet to be reported.

More than 17 million tons of OPEFB is being produced yearly in Malaysia alone [3]. As OPEFB contains more than 50 % (w/w) of convertible sugar polymers (cellulose), it can be a promising biomass for the production of alternative energy carrier such as bioethanol/biobutanol. [4]. But the sugar will not be readily available before breaking down the recalcitrant lignin layers of OPEFB.

Therefore, it is necessary to undergo pretreatment process before hydrolysis of the OPEFB to get the fermentable sugars. Rashid et al [5], Ishmael et al [6] and Amani et al [7] already optimized the physico-chemical, enzymatic and biological processes for the pretreatment of OPEFB. Therefore, journey is going on to look for a sustainable process condition for pretreatment of OPEFB as none of these processes are proven to be either environmentally sustainable.

Enzymes catalyzes many biochemical and chemical reactions and are universally present in plants and animals. Other than the different industrial applications it is highly potential in waste management especially for solid wastes treatment and waste water purification [8]. However, all these desirable characteristics of enzymes and their widespread industrial applications are often obstructing by their lack of long-term operational stability, shelf life and by their recovery & reusability. Enzyme immobilization is one of the strategies to overcome these problems [9, 10]. Therefore, this study is aimed to provide an alternative process for pretreatment of biomasses. Thus in the current research, feasibility study of lignin hydrolytic enzyme immobilization on ferromagnetic particle was conducted to enhance the fermentable sugar recovery rate from OPEFB.

2. Materials and Methods

2.1. Ferromagnetic particles preparation and Encapsulation

Nickel and iron were collected from commercial sources. Then the size of the nickel and iron were determined by using machine Zetasizer Nano S90. The size of the nickel and iron was 2.42 μm, 1.64 μm respectively and laccase enzyme was immobilized on those nanoparticles. Ferromagnetic particles were encapsulated with TEOS using ethanol-water solution at room temperature.

2.2. Characterization of encapsulated ferromagnetic particles

The ferromagnetic particles and encapsulated ferromagnetic particles were characterized using XRD and FTIR. The samples were characterized for the X-ray diffraction patterns, using a Cu K α radiation (λ = 0.154 nm) at 100 mA and 50 Kv. To perform the FTIR, Perkin Elmer spectrum 100 model was used in the transmission mode with wavenumber range of 4000-700 cm^{-1}.

2.3 Immobilization of laccase enzyme on activated ferromagnetic particles and Characterization

The laccase enzyme was brought from sigm-aldrich (Novozym 51003). This laccase enzyme was immobilized on the ferromagentic nanoparticles using the method described by Wang et al [11] with slightly modification. The total protein content of the immobilized enzymes on the ferromagnetic particles was estimated using the Bradford protein assay. The laccase enzyme activity was estimated by the method described by Shah et al [4]. The immobilized laccase enzyme on the ferromagnetic particles was characterized using FTIR.

2.4 Efficacy of immobilized laccase enzyme on OPEFB

The OPEFB was pretreated according to the method described by Shah et al [4] with slightly modification and the degree of pretreatment was assessed using weight loss method. Morphology of the pretreated OPEFB was determined by Scanning Electron Microscope (SEM). The reducing sugar content was determined by Dinitrosalicylic acid (DNS) method of Miller.

3. Results and Discusion

The biomass of OPEFB was collected from Dominion Square Sdn. Bhd, Gambang, Pahang, Malaysia and cellulose, hemicellulose, lignin and ash content was found 43%, 39%, 17% and 1% respectively. Several researchers also reported the similar composition of OPEFB [6, 12].

3.1 Encapsulation of Ferromagnetic Particles

Ferromagnetic particles (Fe/Ni) encapsulation was carried out using mixture of TEOS and $NaOH/NH_4OH$. It was observed that white hollow particles solid were produced by mixture TEOS and NH_4OH while a dense thin white layer was produced by mixture of TEOS and NaOH. Yamashita et al. [13] also reported that no silica particles were formed when using NaOH concentration of 1 to 3M.

Formation of white hollow particles indicates a successful silica encapsulation. Therefore, 5% of NH4OH was selected for encapsulation process, in this study.

3.2 Characterization of Encapsulated Ferromagnetic Particles by XRD

Presence of silica and iron can be detected through the peak at 2-theta values of 44.97° and 65.24° (Figure 1 (a)). Besides, the presence of hematite or iron (III) oxide (Fe_2O_3) contributes to the peak at 2-theta values of 22.86°. Presence of silica can be detected through the peak at 2-theta values of 37.52°, 43.44° while nickel contributes to the peaks at 2-theta values of 44.66° and 51.99° figure 1 (b). Therefore, it can be assumed from this XRD analysis that iron and nickel particles were coated/ encapsulated by silica.

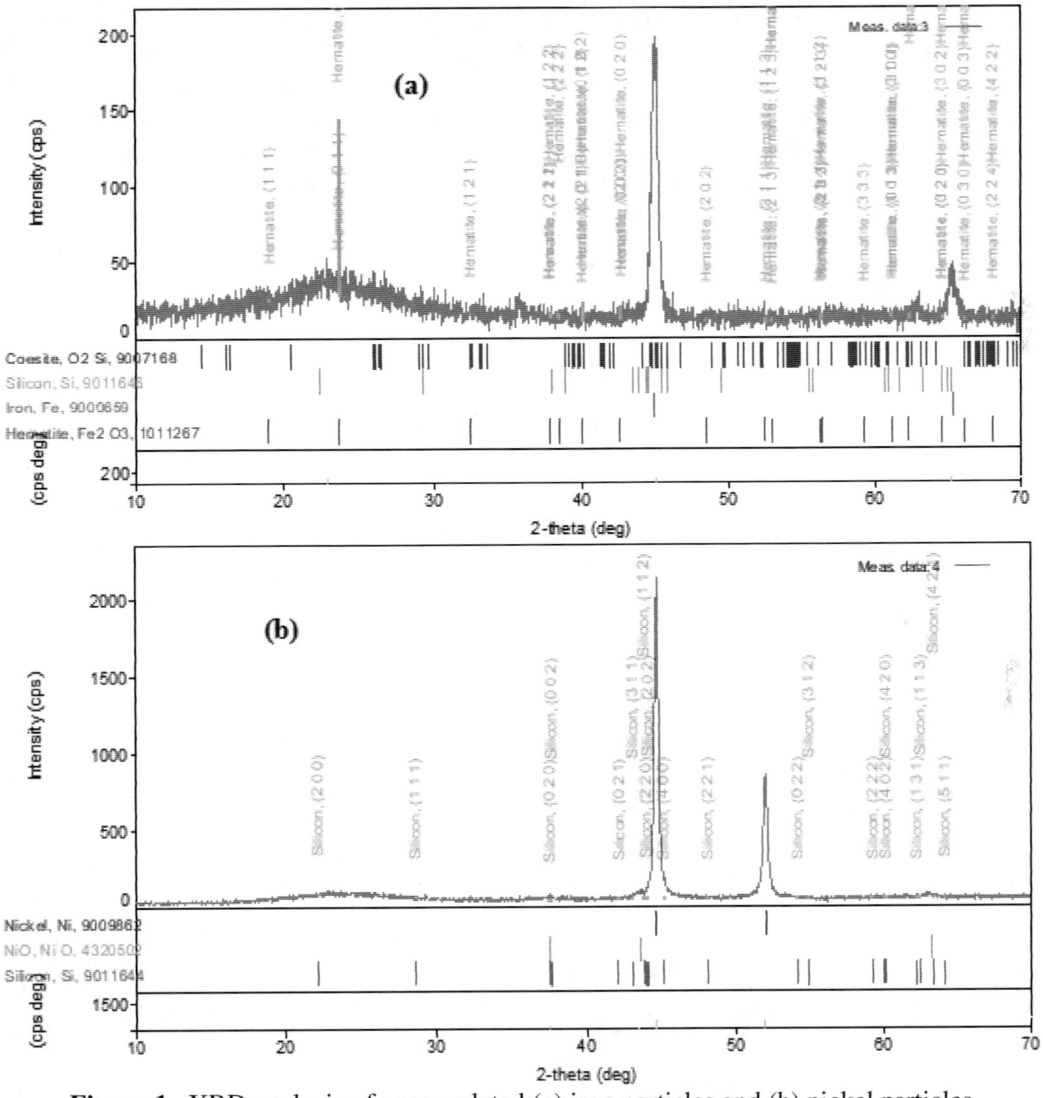

Figure 1. XRD analysis of encapsulated (a) iron particles and (b) nickel particles

3.3 Characterization of Activated Encapsulated Ferromagnetic Particles by FTIR

After functionalization and activation of the encapsulated ferromagnetic particles (Fe/Ni), it was characterized by FTIR and the results were presented by figure 2 (a-b). Silica coating on the iron particles can be recognized by the peak at 1644 cm^{-1} corresponds to Si-OH group and the peak at 1091 cm^{-1} corresponds to Si-O-Si group (data not shown). Cristiano et al. [14] also reported that silica coated magnetic nanoparticles gave a peak at 1616 cm^{-1} corresponds to Si-OH groups and another peak at 1037 cm^{-1} corresponds to Si-O-Si group. FTIR spectrum of immobilized laccase on the activated iron particles (Figure 2 (a)) shows a large increase in the intensity for the peak at around 3444 cm^{-1} which corresponds to -OH group and -NH group contributed by laccase enzyme. Also, there is an increase of intensity for the peak at around 1600 cm^{-1} which corresponds to carbonyl group (C=O) of carboxylic acid. El-Batal et al. [15] also reported the FTIR spectrum for laccase has a peak at 3089 cm^{-1} corresponding to -OH and -NH functional groups and another peak at 1635cm^{-1} which corresponds to carbonyl group. Hence, it could be assumed that laccase enzymes immobilized on the activated iron particles.

Figure 2 (b) represents the FTIR spectrum of activated nickel particles and Si-Ni particle with immobilized laccase enzyme. The characteristic peak at 1634 cm-1 corresponds to Si-OH group and the peak at 1098 cm-1 corresponds to Si-O-Si group. Also, C=N group is present which contributes to the peak at 1400 cm-1. FTIR spectrum of immobilized laccase on the activated nickel particles (Figure 2 (b)) shows that there is a large increase in the intensity for the peak at around 3433 cm-1 which corresponds to -OH group and -NH group contributed by laccase enzyme. Also, there is an increase of the intensity for the peak at around 1598 cm-1 which corresponds to carbonyl group (C=O) of carboxylic acid. Hence, it could be assumed that laccase enzymes immobilized on the activated Si-Ni particles.

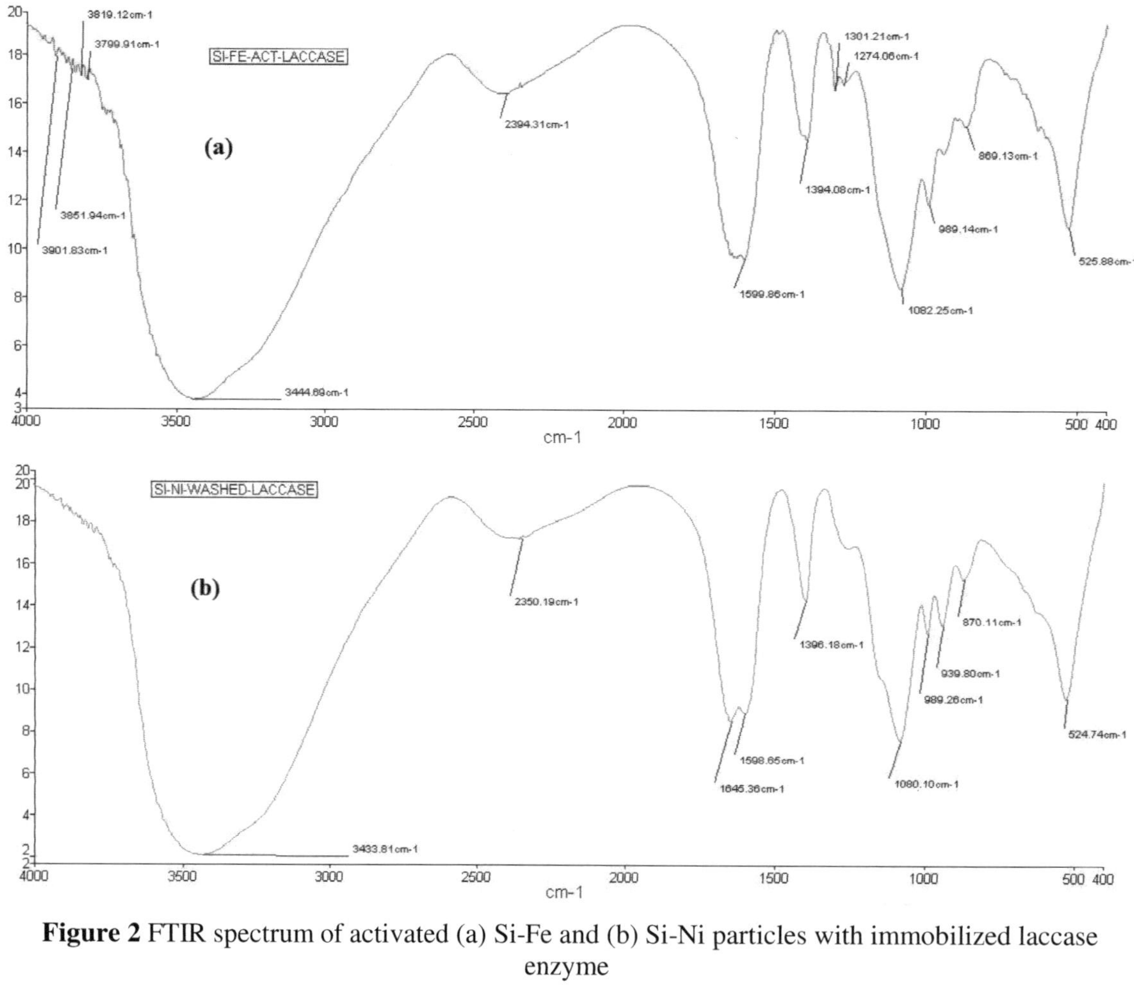

Figure 2 FTIR spectrum of activated (a) Si-Fe and (b) Si-Ni particles with immobilized laccase enzyme

Saccharification of the OPEFB pretreated with the immobilized laccase enzyme was carried out and the efficacy was evaluated based on the released reducing sugar after saccharification. It was found from the current study that saccharification efficiency was doubled when the OPEFB was pretreated with the immobilized laccase enzyme in compare with the un-pretreated OPEFB.

4. Conclusion

Based on the current investigation it could be concluded that pretreatment of OPEFB with laccase enzyme immobilized on the Si-Fe/Ni particle can enhance the saccharification process of OPEFB sharply.

5. References

[1] Littlefield J A, Marriott J, Schivley G A, Skone T J 2017 *J. Clean. Prod.* 148, 118-126
[2] Saba N, Paridah M T, Abdan K 2016 *Ibrahim Construction and Building Materials* 123 15–26
[3] Yoo H M, Park S W, Seo Y C, Kim K H 2019 *J. Env Manag* 234 (2019) 1–7
[4] Shah S R, Ishmael U C, Jaya V P, Asras M F F, Sharifah S 2016 *BioResources* Vol 11, No 2
[5] Rashid S S 2011 Ph.D. thesis submitted to International Islamic University, Malaysia.
[6] Ishmael U C, Shah S R, Jaya V P, Asras M F F, Sharifah S, Victor B A 2016 *BioResources* 11(2), 5013-5032
[7] Amani M H, Makky E A, Shah S R 2017 FGIC 1st Conference on Governance & Integrity Yayasan Pahang, Kuantan, Malaysia, ISBN 978-967-2054-37-5
[8] Rashid S S, Alam M Z, Fazli M B F A 2013 *IJCEBS* Volume 1, Issue 3
[9] Dong M L, Juan C, Yan P S 2018 *Trends in Analytical Chemistry* 102 (2018) 332-342
[10] Bilal M, Iqbal H M N, Hu H, Wang W, Zhang X 2017 *Sci. Total Environ.* 575 1352-1360
[11] Wang F, Guo C, Yang L R, Liu C Z 2010 *Bioresour Technol* 101(23), 8931-8935
[12] Rosli N S, Harun S, Jamaliah M J, Rizafizah O 2017 *Mal J of Ana Sci* Vol 21 No 1: 188 – 196
[13] Yamashita C T, Masafumi A, Hadi R K, Masayoshi F 2019 *Coll and Surfaces A* 566 134–140
[14] Cristiano A B, Silvia R, Luigi M, Mauro B 2018 *The Journal of Supercritical Fluids* 142 1–9
[15] Batal A I E, Nora M E, Aymen S Y, Magdy A A 2015 *Biotechnol Rep (Amst)* Mar; 5: 31–39

ACKNOWLEDGMENT

Authors would like to express their gratitude to the Universiti Malaysia Pahang and the Ministry of Education of Malaysia for the financial support through research grants (PGRS1703102, RDU1703195, RDU190149 and RDU191801-1).

Institute of Physics
Dirac House, Temple Back
Bristol BS1 6BE UK

ISSN: 1757-8981
ISBN 978-1-5108-9831-8
9 781510 898318

CURRAN ASSOCIATES INC.
proceedings.com